LIVING SURFACES

LEONARDO

Seán Cubitt, Editor-in-Chief

Social Media Archeology and Poetics, edited by Judy Malloy, 2016

Practicable: From Participation to Interaction in Contemporary Art, edited by Samuel Bianchini and Erik Verhagen, 2016

Machine Art in the Twentieth Century, Andreas Broeckmann, 2016

Here/There: Telepresence, Touch, and Art at the Interface, Kris Paulsen, 2017

Voicetracks: Attuning to Voice in Media and the Arts, Norie Neumark, 2017

Ecstatic Worlds: Media, Utopias, Ecologies, Janine Marchessault, 2017

Interface as Utopia: The Media Art and Activism of Fred Forest, Michael F. Leruth, 2017

Making Sense: Cognition, Computing, Art, and Embodiment, Simon Penny, 2017

Weather as Medium: Toward a Meteorological Art, Janine Randerson, 2018

Laboratory Lifestyles: The Construction of Scientific Fictions, edited by Sandra Kaji-O'Grady, Chris L. Smith, and Russell Hughes, 2018

Invisible Colors: The Arts of the Atomic Age, Gabrielle Decamous, 2018

Virtual Menageries: Animals as Mediators in Network Cultures, Jody Berland, 2019

From Fingers to Digits: An Artificial Aesthetic, Ernest Edmonds and Margaret A. Boden, 2019

Material Witness: Media, Forensics, Evidence, Susan Schuppli, 2020

Tactics of Interfacing: Encoding Affect in Art and Technology, Ksenia Fedorova, 2020

Giving Bodies Back to Data: Image-Makers, Bricolage, and Reinvention in Magnetic Resonance Technology, Silvia Casini, 2021

A Biography of the Pixel, Alvy Ray Smith, 2021

Living Books: Experiments in the Posthumanities, Janneke Adema, 2021

Art in the Age of Learning Machines, Sofian Audry, 2021

Design in Motion: Film Experiments at the Bauhaus, Laura Frahm, 2022

Northern Sparks: Innovation, Technology and the Arts in Canada from Expo '67 to the Internet Age, Michael Century, 2022

The Artwork as a Living System: Christa Sommerer and Laurent Mignonneau, 1991–2022, edited by Karin Ohlenschläger, Peter Weibel, and Alfred Weidinger, 2023

Computational Formalism: Art History and Machine Learning, Amanda Wasielewski, 2023

Picture Research: The Work of Intermediation from Pre-Photography to Post-Digitization, Nina Lager Vestberg, 2023

Art + DIY Electronics, Garnet Hertz, 2023

Voidopolis, Kat Mustatea, 2023

After Eating: Metabolizing the Arts, Lindsay Kelly, 2023

Tactical Publishing: Using Senses, Software, and Archives in the 21st Century, Alessandro Ludovico, 2023

The Future is Present: Art, Technology, and the Work of Mobile Image, Philip Glahn and Cary Levine, 2023

Living Surfaces: Images, Plants, and Environments of Media, Abelardo Gil-Fournier and Jussi Parikka, 2024

See http://mitpress.mit.edu for a complete list of titles in this series.

LIVING SURFACES

IMAGES, PLANTS, AND ENVIRONMENTS OF MEDIA

ABELARDO GIL-FOURNIER
AND JUSSI PARIKKA

THE MIT PRESS
CAMBRIDGE, MASSACHUSETTS
LONDON, ENGLAND

The MIT Press would like to thank the anonymous peer reviewers who provided comments on drafts of this book. The generous work of academic experts is essential for establishing the authority and quality of our publications. We acknowledge with gratitude the contributions of these otherwise uncredited readers.

This book has been supported by the Czech Science Foundation funded project 19-26865X "Operational Images and Visual Culture: Media Archaeological Investigations."

This book was set in Arnhem Pro and Frank New by Westchester Publishing Services. Printed and bound in the United States of America.

Library of Congress Cataloging-in-Publication Data

Names: Gil-Fournier, Abelardo, author. | Parikka, Jussi, 1976– author.
Title: Living surfaces : images, plants, and environments of media / Abelardo Gil-Fournier, Jussi Parikka.
Description: Cambridge, Massachusetts : The MIT Press, 2024. | Series: Leonardo | Includes bibliographical references and index.
Identifiers: LCCN 2023034668 (print) | LCCN 2023034669 (ebook) | ISBN 9780262547956 (paperback) | ISBN 9780262378475 (epub) | ISBN 9780262378468 (pdf)
Subjects: LCSH: Environmental monitoring—Remote sensing. | Photography in environmental monitoring.
Classification: LCC GE45.R44 G55 2024 (print) | LCC GE45.R44 (ebook) | DDC 580.72/3—dc23/eng/20231025
LC record available at https://lccn.loc.gov/2023034668
LC ebook record available at https://lccn.loc.gov/2023034669

10 9 8 7 6 5 4 3 2 1

CONTENTS

SERIES FOREWORD

Leonardo/The International Society for the Arts, Sciences and Technology fosters transformation at the nexus of art, science, and technology because complex problems require creative solutions. The Leonardo Book Series shares these aims of artistic and scientific experimentation, and publishes books to define problems and discover solutions, to critique old knowledge and create the new.

In the early twentieth century, the arts and sciences seemed to interact instinctively. Modern art and modern poetry were automatically associated with relativity and quantum physics, as if the two were expressions of a single Zeitgeist. At the end of the Second World War, once again it seemed perfectly clear that avant-garde artists, architects, and social planners would join cyberneticists and information theorists to address the problems of the new world order and to create new ways of depicting and understanding its complexity through shared experiences of elegance and experiment. Throughout the twentieth century, the modern constantly mixed art and science.

In the twenty-first century, though, we are no longer modern but contemporary, and now the wedge between art and science that C. P. Snow saw emerging in the 1950s has turned into a culture war. Governments prefer science to arts education, yet stand accused of ignoring or manipulating science. The arts struggle to justify themselves in terms of economic or communicative efficiency that devalues their highest aspirations. And yet never before have artists, scientists, and technologists worked together so closely to create individual and collective works of cultural power and intellectual grace. Leonardo looks beyond predicting dangers and challenges, beyond even planning for the unpredictable. The series publishes books that are both timely and of enduring value—books that address the perils of our time, while also exploring new forms of beauty and understanding.

Seán Cubitt
Editor-in-Chief, Leonardo Book Series

Roger F. Malina
Executive Editor, Leonardo

INTRODUCTION

A giant oak is moving across the surface of the sea. A set of ropes ties branches to the floating platform, while a wooden root ball keeps hold of a cut of the soil with the root system and its microbial environment. A bizarre sight also featured in Salomé Jashi's documentary *Taming the Garden* (2021): an uprooted tree travels on a barge, launched from the Georgian coastline, on its way to the Shekvetili Dendrological Park (figure 0.1). The park advertises that it includes "200 gigantic samples of 39 unique species" of trees, transported from across Georgia—and beyond—in a spectacle of peculiar logistics, captured in images of barges transporting the samples. Part of billionaire Bidzina Ivanishvili's plan, the park seems like a dark counterpoint to afforestation strategies in other parts of the world; the other side of the coin of the plantation logic: "For hundreds of years, these ancient trees have been cultivating their unique underground ecosystems, including vast fungal networks supporting the tree since it was a tiny sapling," said Dr. Kiers, who has been researching how trees are connected. "When a tree is uprooted, that life support system gets ripped from the soil, leaving behind a barren wasteland."[1]

A similar sight: a boat floats a group of trees. This time the backdrop is the New York cityscape as the trees move down the Hudson River (figure 0.2). The *Floating Island* accomplished a previously unrealized project based on artist Robert Smithson's original idea and sketches: "Floating Island is a 30 × 90-foot barge landscaped with earth, rocks, and native trees and shrubs,"[2] making its way around Manhattan. The full name of the project was *Floating Island to Travel around the Island of Manhattan.* The island was meant to act as the "nonsite" double of Central Park, itself an artifice of replaced trees and nature designed by Frederick Law Olmsted, which relied on global flows of matter, especially fertilizers, such as guano from Peru, that transformed the soil.[3] While Smithson was interested in designing an artificial island, he already saw Central Park as an island located on land: "A park can no longer be seen as 'a-thing-in-itself,' but rather as a process of ongoing relationships existing in a physical region."[4] The original idea for the floating tree island from the early 1970s was introduced in the context of Smithson's interest in large-scale earthworks, where even the proposal (pencil drawings) was considered an artwork in diagrammatic form.[5] (figure 0.3) The actual ship and trees on the river in 2005 were, then, a double of a double of the park, an artifice upon an artifice. A drawing upon a diagram represents a transfiguration of vegetal forms and a paper form into a living sculpture.

Some five hundred years earlier, a similar scene was drawn by Leonardo da Vinci in *An Allegory with a Dog and an Eagle* (figure 0.4), dating from circa 1508–1510. A dog—sometimes interpreted as a wolf—is steering a sailing ship with a mast of a tree, bearing striking visual resonances to our more contemporary examples of uprooted trees on waterways. The interpretations of Leonardo's drawing include varying suggestions that it is an allegory of European geopolitical alliances involving, for example, France, Pope Leo X, the Holy Roman Empire, or "even . . . of canalisation projects in Lombardy."[6] We are not going to attempt to provide another competing interpretation as we are focused on the visual theme that is captured, for example, as part of one of the panels in Aby Warburg's *Mnemosyne Atlas*.[7] To extend the Warburgian method, we juxtapose these scenes of trees on seas and rivers, and ask what sort of themes they start to form. They can be read like a media archaeological topos, a recurring visual or discursive theme that gets reactivated in different historical and social contexts, prompting a reading that moves forward and backward: the sixteenth-century allegory read in relation to the Georgian tree park and its geopolitics, or Smithson's artistic note on site and nonsite as part of designed natures that work on visual themes, familiar since the Renaissance.[8] Following this recurring theme, we ask: What

0.1

Taming the Garden by Salomé Jashi, Mira Film / Corso Film / Sakdoc Film (2021). Film still used with permission.

0.2

Robert Smithson, *Floating Island to Travel around Manhattan Island* (1970/2005). Produced by Minetta Brook in collaboration with the Whitney Museum of American Art. On view September 17–25, 2005. Photograph: Andrew Cross © Holt/Smithson Foundation / Licensed by Artists Rights Society, New York.

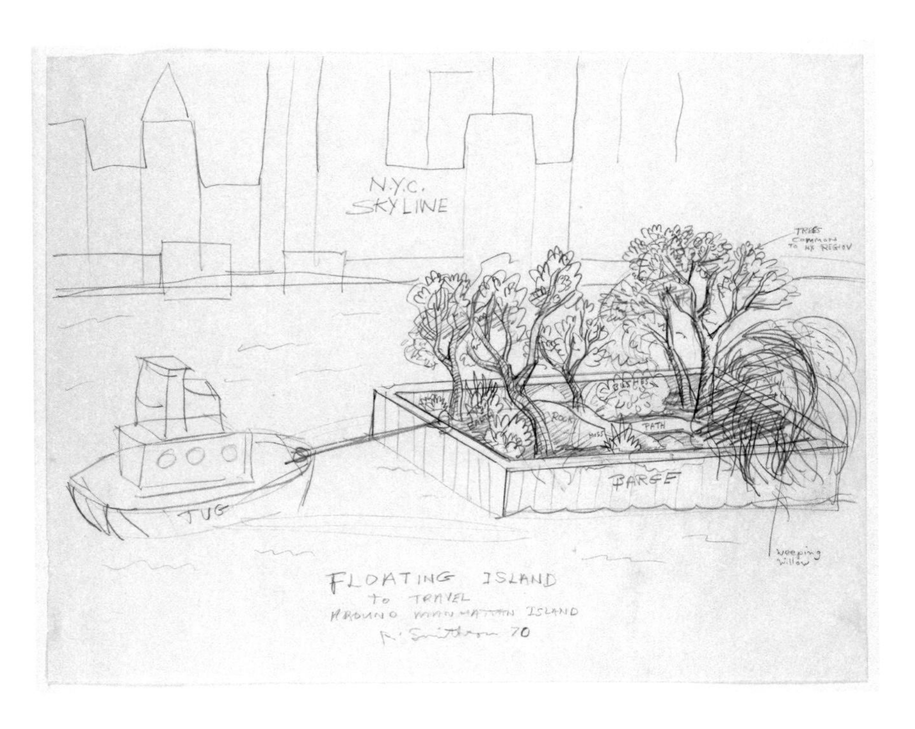

0.3

Robert Smithson, *Floating Island to Travel around Manhattan Island* (1970). Graphite on paper. 19×24 in. (48.3×61 cm).

0.4

Leonardo da Vinci, *An Allegory with a Dog and an Eagle* (ca. 1508–1510).

sort of an image of design emerges if we start with this recurring image of transported trees on barges? How does this historical topos illuminate contemporary concerns with shifting landmasses and islands, the aesthetics of planetary surfaces, and the art and design practices that respond to them?

What we read in and through the three scenes (or four if you count Smithson's drawing as separate from the realization) does not concern only author-based practices of landscape design or critical gardening. This topos and design start to concern *medium design* instead, to echo Keller Easterling: we must attend to "protocols of interplay," immanent potentials of assemblages, active forms embedded in all sorts of material constellations between things.[9] The logistical theme of uprooted trees and the transport of plants and other vegetation isolated from the soil for such displacement and transport is a central topos of a more recent situation concerning living vegetal surfaces and the planetary condition. The images and their stories speak of synthetic natures where the natural and the artificial have merged in a broader context of landscape transformations that consist of shifting landmasses, transported trees, and mass projects of deforestation and afforestation.

Shift from the figure to the ground, though, to what is not necessarily visible in the first instance. The background is as important as the ships carrying trees—the elements mirror each other. Water is the floating medium for traveling islands, while the images depict interfaces: this series draws relations between air, water, life, and the vast surface and subsurface abyss of seas and oceans that are themselves lifted to allegories in many of our contemporary discourses about planetary biodiversity and climate. Such an extended allegory has to however take into account what is not immediately evident in the images we started with: while water surfaces as a thin layer only, it unfolds as depths of marine life. The richness of any contemporary allegory would have to include nonhuman life of animals and habits, and the movement of entire surfaces that are the fundamental life-support systems we are concerned with. Such depths might not be so evident in any of the examples we started with, but they do become more and more relevant as we investigate surfaces as access points to the wider logistics of life.

Among such a focus on logistics of life, we need to take into account the logistics of the image as well. While water, the liquid medium, enables the islands to glide, it surrounds and isolates them as if they were the figure of an image of sorts. Water and visual depictions converge, and even more, image and logistics, and images of logistics too, as the movement of living trees becomes entangled in the continuous stream of circulating images that register movement as well as light. Two kinds of light merge: registering light through photosynthesis and registering light as it forms an image.

Such cases of material shape-shifting, to echo Jane Hutton, intrigue us. While we share an interest in maps, diagrams, and images of materials and landscapes that drove Smithson, light—beyond images—acts as one agent in the transformation of planetary surfaces in how growth is managed, regulated, accelerated, or decelerated.[10] If "speeding up and slowing down landscape processes is the design challenge *of the times*,"[11] we have to also account for the light, images, and other energetic mediations that take part in this environmental drift.[12]

This book is not about symbolic allegories but nature in movement and the living surfaces that define the planet as a material, chemical, and visual inscription. We examine the hypothesis that the planet consists of surfaces and subsequent interfaces that function as images and images embedded on planetary surfaces as part of complex observation, monitoring, (remote) sensing, and imaging techniques. *Living Surfaces* addresses how the surface of the earth has, over the last two centuries, become known and perceived as an environment of images. In other words, the understanding that the planet is made of material surfaces of different types is linked to the production and distribution of the synthetic surfaces of images. These can be formed on the visual light spectrum or radiating outside it, circulating as paper maps and geographic information system (GIS) data layers, pixel-related landcover indexes, and image-based datasets of contemporary environmental monitoring while aggregated and streamed on commercial and non-commercial data platforms.

More specifically, our argument builds on the history of observation of vegetal phenomena through visual media, such as photography, (early) moving images, and recent digital technologies employing massive datasets of remote sensing. The historical examples also imply a broader argument about ecological aesthetics and media theory. By focusing on images and imaging to understand the growth and movements of the surface of the earth, we demonstrate how this link is both historically grounded and fruitful as a proposal about the aesthetic and scientific operations of images. In other words, this entwining develops not only into specific cases of management of the environmental surface but also gives rise to an ecological aesthetics of media.[13]

The proposed ecological aesthetic deals with different spatiotemporal scales of vegetal life from the geode to the plant, from the agricultural plot to laboratory situations of observing growth. The book offers a multiscalar cartography of living surfaces and vegetal images—not so much about images of plants but plant life as it is entangled with practices of light, imaging, and modeling—that helps us to understand core questions that pertain to current discussions about the Anthropocene. There exists a significant art history of botany (in part also about the history of colonialism);[14] an entire media history of photography of plants—and photography *by* plants, as Harold Wager claimed in 1909;[15] and a cinematic history of how slow-motion and time-lapse techniques participated in the epistemic unfolding of plant growth and movement.[16] Of course, one could add multiple layers to this art- and media-infused narrative that would speak to issues of cybernetics and data as the later knowledge techniques of planetary plant surfaces, but this is why we wrote this book in the first place. The questions of what grows, how to accelerate growth, and how to sustain planetary ecological diversity are central to environmental, agricultural, and vegetal life as it is entangled with the planetary scale issues of climate change and the biodiversity crisis. We are, however, trying to find a media theoretical angle that discusses this from a particular historical and contemporary perspective. Such an argument seems to merge seemingly separate domains, from geography to biology to visual culture studies. Our take is more specific than this sweeping note first seems. We are after a particular genealogy of the entanglement of technical imaging with vegetal surfaces that makes sense of how the surface of the earth is changing and how practices of images—and the broader cultural techniques of light—inform the environmental outlook that has taken central importance over the past decades of scientific and nonscientific knowledge about climate change and the parallel plethora of damage to the biodiverse planet. While images of uprooted trees serve as one entry to our logistical interests, we will elaborate on what we mean by the centrality of surfaces as aesthetic and epistemic units with lives of their own.

ON THE READABLE SURFACE OF THINGS

Life on earth occurs on the surface. A thin film on the uppermost crust of the globe receives energy from the sun and transforms it into a layer that contains all living forms of the planet. Since the beginning of the twentieth century, this layer has been identified and observed as a continuous entity defined by biochemical cycles and recursive loops operating at the scale of the geode. This sphere that wraps around the planet has received different names—such as the biosphere or the critical zone—and its study as an ecological platform lies at the core of the theory of Gaia as well as other approaches in which the earth is conceptualized as a self-regulated homeostatic body.[17] Life, observed at the massive volumetric scale of the body of the planet, is flattened onto an animated envelope. From such a point of view, living films of photosynthetic matter evoke other films of photosensitive movement: we could call this material shape-shifting, echoing Hutton again. As we argue, this shape-shifting appeared even before satellite images of the earth made this point of view explicit and organized into time-lapse series. Already glass spheres in laboratories, outdoor photometric measurements, week-long time-lapses, or hemispherical photography merged living surfaces with the surfaces of images, presenting a material and epistemic aesthetic for images of growth.

Not by chance, this dynamic body is often seen as a "green mantle," driven largely by the transformative capacity of chlorophyll to "absorb the radiant energy of the sun and utilize it in bringing about certain chemical changes inside the plant as the result of which sugars, starch, fats and oils, and proteins are formed from simple substances present in the soil and in the atmosphere," as one popular take on the elemental media of "plantscapes" put it in 1939.[18] The greenness of the green mantle, however, is not merely the residual color after photosynthesis, taking place at the scale of the planet, but something that has become registered in remote sensing, in analytical classifications, as well as in different landscape and landcover maps, alongside the range of the other color codings embedded in elevation maps: a few shades of green between the high white peaks of mountains and the deep blue of ocean depths are linked to an epistemology of geographic information and data.[19] The green mantle, or the living film, marks thus a historically situated surface epistemology and aesthetics, from the farming of the vast prairies in the American Midwest in the early twentieth century to the large-scale landscape projects such as the Great Green Wall of the Sahara or the irrigation agriculture in the Gobi Desert.

In addition to color maps as one (data) visualization of the earth's surface, other information systems built on the surface of the image (photomosaics, layer models, GIS) register the existence of different types of regions, such as ecozones, biomes, ecosystems, and biotopes. Such surfaces are not traditionally considered natural, but they index different scales of change and transformation of the mantle as a living formation. The surface and atmosphere are also built environments, sometimes captured in terms such as the *archaeosphere*. In Matt Edgeworth's words, the archaeosphere is "a giant carpet covering large areas, on which the furniture of the human world (its buildings, monuments, pylons, oil-rigs, telegraph poles, roads, railway viaducts, shanty towns, parks, airports) stands and is supported, and into which it will eventually crumble. Deep layered in places, threadbare and patchy in others, this carpet of near-global extent provides the surface on which people carry out their lives."[20]

Similarly, a multitude of other transformations are part of this technological condition of inscription that has radically shaped the earth: large-scale mining, deforestation, afforestation, plantation projects, and other complex sets of medium design that challenge what we understand as surface as well as its scale, involving different simultaneous cultural techniques of terraforming of the planetary surface.[21]

The persisting overlap of the biosphere and the technosphere—or medianatures[22]—is one example of how this surface-carpet continuum is not merely one single object of a planetary body; instead, the surface of living matter appears as a multiscalar composite of multiple patches, levels, and textures, displaying different arrangements of flows of matter and energy. Flows and processes, taken as a whole, express the complexity of the surface as our opening series of images also exemplified. Without a privileged scale, the environmental surface can be seen as a relational composition, a spatial distribution of organization of matter formed as the result of continuous flux of radiation from outer space. The history of the biosphere and investigations of critical zones point to the thin layer of life on the planetary crust, which can be paired with our technologically mediated knowledge of life, growth, energy, and more that also takes place in and through surfaces. Sensing and imaging living surfaces come to transform those surfaces that are increasingly depicted as images in their own right.

Throughout the book we return to instances of how surfaces and images are intimately related. We consider how the surface as an interface of multiple composite forces—mostly related to plants entwined with light—has become an interesting case for imaging and modeling change in microscopic or geographic domains in urban and nonurban situations. "Landscapes are models *in situ*," according to Jane Hutton.[23] Vice versa, the chemically mediated changes produced at the level of the soil or the plant are echoed in image practices that link up with the material transformations, whether this concerns laboratory experiments, outdoor methods for measuring plant formations in prairies, or computer simulations of grasslands.

Since the era of mechanically assisted flight, images have been essential for witnessing and tracking the changes in land cover. This concerns specifically photographic images, especially the aerial images taken from airplanes and later by satellites, including also the nonphotographic forms of remote sensing, such as multi- and hyperspectral imaging, that expand sensing beyond the visible frequencies of radiation. Emerging from sensing and visualization, different operational diagrams and mapping systems forming "geo-epistemic" practices have also directly manipulated geographic territories.[24] This has been central to many earth observation systems. Remote sensing technologies in the search for geographic features revealing oil deposits would be one example; another would be monitoring crops' health to predict yields and to preemptively intervene to protect or enhance growth.[25]

In short, while different anthropogenic changes, including climate change, can be seen as a rewriting of the earth's surfaces (and atmospheres), as Kathryn Yusoff argues, techniques of reading surfaces as images come to claim a central place in making sense of the planetary.[26] These are not only images in the photographic sense but arise from a chain of techniques of surveying, monitoring, and remote and near sensing that visualize even subsurfaces (e.g., in archaeology, as well as in extraction industries, or under the surface of water, as in ocean floor mining), as Geoff Manaugh argues, employing the term "geomedia."[27] Such environmental surfaces and the realm of geomedia are also sites of military operations, postcolonial conflicts, environmental degradations, and persisting territorial claims."[28] This theme is evidenced in other cases of violent medium design too. Herbicidal warfare repainted the planetary surface colors through synthetic chemistry. Echoing Harun Farocki's point about industrial production and destruction working side by side, fertilizers and defoliators are chemicals that accelerate or decelerate growth: we invite you to see these in parallel terms as the merger of surface, light, and film. They are one variation of the story of the green revolution as a particular theme of political ecology and aesthetics that also has one of its arms in warfare and colonial conflicts.

THE SURFACE AS SITE OF MATERIAL RELATIONS

While the surfaces we deal with are quite specifically about plants and environmental imaging, including inscriptions and material changes from soil chemistry to geographical extensions, we are building on existing traditions of what the surface has meant in media studies and beyond. A quick recap is thus in order to point out the divergences as well as convergences. A significant part of critical theory and cultural criticism of the early twentieth century was focused on the surface in one way or another. They include Siegfried Kracauer's reading of early media technological urban modernity in terms of its surfaces, although it often has an implicit melancholy regarding the loss of depth and meaning. Beyond this, Kracauer's notes on movement, photography, film, industrialization, and different architectural surfaces were underpinned by his take on "surface-level expressions," articulating the surface itself as a form of auto-inscription of cultural change: "The position that an epoch occupies in the historical process can be determined more strikingly from an analysis of its inconspicuous surface-level expressions than from that epoch's judgments about itself."[29] Kracauer's take on the "unconscious nature" of such expressions led to some of the most referenced writings of modern cultural theory, while the surface itself was left less investigated in its own right.

The surface returns in other media, art, and architectural writings, while its relation to flatness became a core topic of discussions in aesthetics. Voiced soon after the Second World War, Clement Greenberg's claim that the "world was stripped of its surface, of its skin, and the skin was spread flat on the flatness of the picture plane"[30] was meant to mark the separation of modernist aesthetics from earlier legacies of tactility. Here, though, Greenberg's reading of Cubism appears as an operation to prepare flatness so that "we no longer peer through the object-surface into what is not itself; now the unity and integrity of the visual continuum, as a continuum, supplants tactile nature as the model of the unity and integrity of pictorial space."[31] Yet, surfaces abound with tactile, architectural, textural, and even cinematic depth and motion. The theme and subsequent conceptualization of surface have proven elastic enough to deal with different realms of materiality across a body of disciplines: film and architecture, design, color, industrial chemistry, postmodern simulacra, digital spaces, and the topology of social networks.[32]

Giuliana Bruno's influential work on surfaces aims at the intersection of projection (cinema), architecture (space), and fashion (texture and embodiment). Working against the conflation of surface with flatness, Bruno places questions of texture and (haptic) visuality in close conversation. The surface becomes addressed as sedimented, resonating with media archaeological notions of layers of history specific across material instances, even geographies. Maps and material surfaces interact. Navigation across space is, besides motion, also emotion—the affective involvement of different kinds of bodies in interaction. The textures of surfaces are related to practices of light, such as projection—including experimental contemporary art projects—as well as to an extended understanding of the screen object; architectural facades and garments feature much outside the standardized format of a screen to watch; surfaces give birth to textures and atmospheres in various (primarily) urban situations.[33]

Among a discussion of various projects that concern the new modern materiality of (electric) light, Bruno picks up on László Moholy-Nagy's work as an example of how visual arts engage with modern surfaces such as the screen. A whole aesthetic repertoire emerges from the experimental arts-architectural spectrum of the early twentieth century. For Bruno, Moholy-Nagy's focus on surfaces as sites of experimentation connects with the realization that the screen is "an actual surface, . . . a material in itself."[34] While the condensation of material environments onto (cinematic) screens is one theme that runs through Bruno's analysis, it also shifts to the architectural inversion where the atmospheric, environmental notions of cinema are designed in space.[35] Bruno emphasizes that the screen is thus, fundamentally, a landscape.

For us, the notion of the surface becomes a way to discuss the proximity and merger of vegetal surfaces with media and image surfaces. Understood this way, it scales beyond buildings or cities, beyond human bodies and experience. It points to a different genealogy of scientific and technological experimentation, where questions of plants, (nonhuman) biological growth, and planetarity come to the fore, from leaf surfaces to atmospheric chemistry to the architecture and the chemical basis of contemporary atmopolitics[36] (figure 0.5). This pushes the logic of experimental practices *a la* Moholy-Nagy further. If screens are landscapes, we are interested in the inversion of landscapes as screens become picked up in epistemic contexts, from labs to plant geography to remote sensing.

In other words, we draw on recent methodological and thematic entries on the topic of the surface, while our case studies extend to the media ecological dynamics captured in situations of vegetal growth, exchange of energy, and the production of data. The surface functions as an epistemic unit that we map historically from plant cells to agricultural colonizations to planetary envelopes. We aim to discuss the shift from being "unconscious" of the surface to seeing in it a broader repertoire of forces—including biological, chemical, and geophysical. Such an approach expands the interest from the urban modern to the wider scales of planetary diagrams and autoinscriptions.[37]

It is here that we also find close affinity with many of the forensic approaches, such as those developed by Eyal Weizman and Susan Schuppli, as they write about the double inscription that becomes a crucial epistemic and aesthetic tool for analysis. What if surfaces are already image-like inscriptions and can unfold stories of conflict, violence, and anthropogenic change? What if any sensing and images register what already was registered across the surface that is both natural and an

0.5

Architectures such as nineteenth-century winter gardens bring into play different elements—surface tensions—from images of the drawn landscapes to the glass housing for the plants inclusive of the water surface. Wintergarten König Ludwig II, ca. 1870. Albert, Joseph (Munich). Public domain.

artifice? What if surfaces such as deserts and forests are to be read like photographic inscriptions, as Eyal Weizman or artists such as Jananne al-Ani suggest?[38] For example, aerial photography is an example of the production of images of existing inscriptions: "the subtle traces of what has been erased: traces of ruined homes and small agricultural installations, of fields and wells that can sometimes be noticed under the grid of newly planted forests, as well as the dark stains of long-removed livestock pens."[39] Hence, the surface becomes a methodological part of media-driven forensic investigations, and that emphasizes that our interest in logistics of images and movement is not primarily allegorical.

ENVIRONMENTAL MEDIATIONS

The merger of surfaces and images is a key driver of the argument of this book. The two terms start to define each other and take different roles, whether as an object of research or as its methodological scaffolding: to read histories, layers, violence, and conflicts from surfaces. Our interest in the surface, as noted before, is due to its particular epistemic role as a "site": on the one hand, a formation through which planetary processes have become known; on the other hand, a constantly fabricated element—sometimes a landscape, sometimes a plantation plot—that itself takes on the characteristics of an image (grids, frames, color codings, inscriptions and pattern recognition, change and movement, etc.). As such, this could be seen as a peculiar entry point to ecomedia at the scale of the planet: if cinema and other technical media are cosmomorphic, to use Adrian Ivakhiv's coinage, in how they fabricate "the shape of . . . a world, a cosmos of subjects and objects, actors and situations, figures moving and the grounds they move upon," then so are the material dynamics out of which a peculiar sort of imaging appear as inscriptions, movements, chemical transformations, atmospheres, and formatting of different figures of biological—and especially vegetal—surfaces.[40]

We claim that the logistics of moving vegetation (like uprooted trees discussed above) characterizes much of contemporary ecomedia: not just moving images that depict ecological and climatological change, but ecology itself as the movement of matter and images. Not all of this is as clear-cut as an image as the barge with a tree and its soil subsystem on top of water surfaces, and some of it concerns the transformations of media involved. In other words, as Graig Uhlin argues, film as a "pulverized plant, the organic made technological," carries forward the processes of photosynthesis as it becomes an image.[41] Such images are one example of crystallization of cosmic energy or, in more mundane terms, the period of fossil fuel dependency. This cosmic energy then also became part of cinema, and other practices of light, as exemplified in Nadia Bozak's proposal to discuss images as "fossilized sun."[42] This resonates so closely with much biosphere thinking that emerged around the same period as early cinema in the late nineteenth and early twentieth centuries; we are here thinking of, for example, Vladimir Vernadsky's work on the planetary surface engine of photosynthesis (chapter 3).

Environmental surfaces emerge as multiscalar composites linked to material arrangements of both planetary matter and cultures of imaging. As such, they are interlocked co-composing forces. While environmental surfaces would seem at first firmly bound to geographical areas, a large amount of research in environmental sciences and humanities describes them as dynamic and changing: landscapes migrate, furs and vessels carry their seeds, biomes shrink and grow, and climatic regions shift as part of what landscape architect James Corner refers to as *terra fluxus* instead of a *terra firma*.[43] Even geographical and geological surfaces are radically temporal, and this feature has affected the development and the understanding of how imaging can sufficiently convey such processes that are dynamic and defined by change.[44] The synthetic layers of visual monitoring that feature in photomosaics and tilemaps have given rise to temporal sequences that work with time-lapses and other quasi-cinematic methods in an expanded aesthetics of ecology.[45]

The *terra fluxus* necessitates *imago fluxus*. In Walter Benjamin's words, "To each truly new configuration of nature—and, at bottom, technology is just such a configuration—there correspond new 'images.'"[46] A different regime of practices of light has emerged, which includes the nonvisual spectrum of radiation, for example in contexts of remote sensing or precision agriculture. Light is registered not merely as visualization in the usual sense but also for the acceleration of growth, like in the many experimental research sites such as greenhouses and phytotrons, where, for instance, ultraviolet light has become a tool for stimulating plant growth and aging fruit. Not by chance, while the rapidly changing conditions associated with the Anthropocene have accelerated the pace of some of these vegetal patterns—whether contained inside buildings or out in the open, defining surfaces wrapped as "landcover"—the promoters of the most advanced monitoring techniques market their products as "a living atlas for the world."[47] The ideal of constant real-time monitoring of the earth's surface that emerged in the Cold War period, with satellite sensing, continues in our current period of remote sensing combined with data analytics and prediction techniques of machine learning.

CULTURAL TECHNIQUES

The various strands of environmental media research and practice in art and design have provided conceptual and methodological tools to expand questions of media to their material infrastructures. This expanded notion also includes how environment becomes a condition of mediation while it is simultaneously shaped by techniques of mediation. Relating to the term cosmomorphic, Parikka's notion of medianatures refers to the continuum from natural resources to technological media in material and epistemic recursions: one conditions the other.[48] So-called nature is necessary for contemporary media technological infrastructures to function (as energy, rare earth minerals, and more), while media technological sensing, representations, visualizations, imaging, and circulation frame this so-called nature.[49] In other words, medianatures establish a framework where material affordances are seen as essential for the existence and operation of contemporary technologies. Furthermore, medianatures mark the double movement where media technologies of sensing, visualizing, calculating, addressing, and modeling are also the means through which natural formations, territories, and resources are observed and tracked, and their changes are predicted. This offers further insight into why multiple kinds of vegetal surfaces are to be understood in relation to media technological operations. These operations can cascade into different levels of abstraction, from the concreteness of near-sensing soil to the abstractions of financial modeling of different geographic territories with a view to the potential of extraction industries.[50]

Besides the focus on ecology and media that comes out in different guises in contemporary scholarship, our methodological commitment to the notion of cultural techniques is a guiding line through the chapters. Emerging from certain strands of (so-called) German media theory, cultural techniques have grown from connotations in agricultural engineering and the transformation of nature into a body of work on media techniques. Here techniques are understood as dynamic operations that give rise to the objects that are their referents. Cultural techniques of numbering, spatial divisions, architectures, subject positions, and more are part of a realization that is recursive at its core. They work upon their own conditions: how there are images about images or writing about writing or, in more complex terms, how computational techniques to model and refer to the earth's surfaces come to transform those very surfaces.[51] In Bernhard Siegert's sober description, "The analysis of cultural techniques observes and describes techniques involved in operationalizing distinctions in the real."[52] Cornelia Vismann's example of a plow inscribing a line and delineating a plot is exemplary here: it is both a material event and evidence of the emergence of symbolic inscriptions such as property (as she argues was the case in the *Imperium Romanum*).[53] Such inscriptions of the real can be distinctions, like an inside and outside, but may also refer to the production of other forms of difference that take shape in the material world: some that we deal with in this book include practices of light from photography to photometry, glassware in laboratories and architecture, the controlled growth of irrigation systems, and remote sensing and ground truth, all of which operate at the back of how the earth has been formatted for sensing and modification (chapter 5). Furthermore, some of the cultural techniques we engage with are enacted as concepts like the biosphere: the term for a dynamic realm of biogeochemical interactions that comes to function as part of a cosmic interface (chapter 3), a large-scale differentiation of insides and outsides, of solar energy and terrestrial geopolitics.

Throughout the book, we are interested in how some of these inscriptions come to take the place of the surface itself. That is, when images come to stand in for what they represent, second-order techniques frame what is considered a surface. The production of surfaces is thus one fundamental example of a cultural technique that concerns literal agricultural techniques (including data practices such as precision farming) as much as it does, for example, ground truth markers, plants used as environmental indicators, or other elements along a semiotic-material continuum. Such chains of "inconspicuous cultural techniques"[54] concern themselves with the complex relations between visual and invisual operations.[55] Our interest in knowledge objects that stand at the center of operations concerning remote sensing, quantifying vegetal surfaces, and measurement frames what we think of as the ecology of images: an assemblage of surfaces and technical images that recursively shape each other.

ON THE SURFACE OF PLANTS

The examples of cultural techniques that are most often cited deal with elementary operations such as drawing lines, gridding, or counting, carried on either by humans or by mechanisms designed by humans. This would seem to suggest that the "natural" side of the production of cultural differentiation is ignored—with some exceptions—by this type of approach.[56]

To understand vegetal matter in relation to cultural techniques, let us return for a moment to the opening images of floating islands and the surrounding waters. Water is already a form of visual media, not only for its reflective surface but for the swirls and eddies that define its surface-level patterns. So, what does it mean to draw or to picture water? Such a question emerged also in Leonardo da Vinci's work: in order to learn to draw the swirls of water, he designed a series of simple bowls to keep the swirls stabilized. The bowls included containers with swirl-shaped borders, which made water always reproduce the same swirl. Drawing was thus embedded in experimental systems that framed and reproduced surface patterns by design. To quote Siegert:

> When Leonardo draws water in motion, he is not only drawing its ornamental shapes but also observing design as something that takes place within the experimental system (rather than out there in nature). By virtue of its fluidity and

the innumerable shapes it assumes, water more than anything else resembles imagination—it is a designing, inventive entity. By subjecting water to certain arrangements, the engineer can perpetuate (though not arrest) its swirling motions, while the artist, in turn, is able to wrench an image from this liquid spirit. We do not start with a ground against which the figurations stand off; rather, ornamental and grotesque figurations create the ground from which they, in turn, emerge.[57]

The cultural technique in Leonardo's experiments resides in the skills of observation and drawing as well as in the preparation of the containers for the fluid in movement. Flowing through them, the movements of water become ornamental signatures outlined by the prepared containers and the drawings. That is, the experiment contains the forms that are produced by the water on the surface of the drawing. The swirl pattern, the drawing, and the designed container contribute to the production of the stabilized form that will be identified and repeated afterward. Here, water is the inventive element, but it is through the fabrication of cultural techniques, including material devices, that these powers of invention can be recognized and repeated.

Our interest is in such material devices that appear and shape contingent historical conditions for observing and describing plant life. Instead of drawings of water, our case studies concern cultural techniques that give rise to characterizations, classifications, and applications of plants while at the same time grounding the experimental setups themselves. This is, we claim, one instance of medium design in the manner Easterling defines it and which we quoted earlier. We can even refer to *so-called plants*, as Jara Rocha and Femke Snelting do: they are not taken as instances of any natural category but are parts of a continuous and dynamic reconfiguration with knowledge systems and material practices.[58] As such, the cases in this book revolve around experiments that have dealt with three main ideas that characterize the environmental surface where plants and images converge: environmental sentience, surface inventiveness, and atmospheric transformation.

Regarding *environmental sentience*, the book overlaps arguments about the sentience of vegetal life as a nonhuman form of intelligence which has become an emerging research topic over the past years in Plant Studies.[59]

The influence of the field can be seen in contemporary art and design, including bioart.[60] Focusing on the work of pioneers of photography and educational cinematography, such as Anna Atkins, William Henry Fox Talbot, Oskar Messter, Mary Field, F. Percy Smith, and Margaret Thompson, much of the literature on visual and media histories of plants highlights how photography and film facilitated the framing of questions about the agency of plants.[61] Conversely, other approaches have emphasized how the representation of plants in movement—thanks to the new media of the time-lapse and cinematography—fueled the early twentieth view that media themselves animated with traits of life, a zoetic stance that also feeds into more recent work nonhuman photography and cinema.[62] Our book, however, is not trying to argue about plant sentience or media animism but looks at how vegetal matter is framed, measured, and occasionally (re)produced in experimental assemblages.

In addition to environmental sentience, *surface inventiveness* characterizes plants. The vegetal realm is creative in forms and colors. Such designs respond to material contingencies at the scale of their surroundings—such as viruses affecting the patterns of tulips or the trophic chains they belong to, as those plants that mimic the shape of predators. Plants are "natural sensors" and "in-situ imagers," as Susan Schuppli puts it: they express through their forms and colors the characteristics of their milieus.[63] This aspect explains why plants have featured in many contexts of knowledge production and operational setups. Contemporary precision farming, for example, exploits the information contained in the electromagnetic spectrum of frequencies emitted by plants. When connected to technological circuits of agriculture, plants are taken as broadcasters of information.[64] They inform about their own health as well as about the quality of the soil underneath, adding a signal layer that features as a backstory of planetary communication of energy, chemicals, light, and color as proxies of ecological life.[65] To focus on cultural techniques related to emerging practices such as precision agriculture involves attending to how plants are integrated into the circuits of value production as the techniques of finance, such as rural and farm assets.[66] Here imaging and data, sensing, and information involve many complex recursive techniques that squeeze the plant into the managerial surface, to borrow a term from John May.[67] Plant expressiveness is transferred onto a surface that becomes an interfacial element. It registers data that, in turn, facilitate procedures of management that, in the current moment, also relate to operations of machine learning and prediction.[68]

Finally, in addition to their sentience and inventiveness, *plants transform their surroundings*: they construct environments and atmospheres. This aspect features rhetorically where, for example, forests are metonyms for the ecological equilibrium of the whole planet, but it also has a broader context of meaning.[69] We address how the atmospheres created by plants have been sampled and analyzed through techniques of enclosing and modeling at different scales: from prototypes of vegetal cell membranes to glass bell jars, greenhouses, phytotrons—even to forest light climate measurements and models of planetary envelopes. This recursive nesting of architectures, media, and operations is at the core of how the terraforming ability of plants has been placed onto the domain of data images that form the central component of current environmental surfaces.

Plants, headless entities entwined with their surroundings and expansive acclimatizers of environments, call for radical reformulations of the definitions of and relations between individuals, technologies, communities, and environments.[70] Besides being inspiring processes to think with, their merger with images and the technical visual culture results in alternative medianatural envelopes.

In this regard, the work by Jennifer Gabrys and Paulo Tavares helps us to understand some of those envelopes. Gabrys's research project on smart forests describes and analyzes multiple processes of the datafication of forests that uses networks of remote and ground-level sensors together with other forms of knowledge production. Her focus is on understanding how forests are transformed by data and, reciprocally, how designs of sensor networks are refashioned by forests to help us see different aspects of the forests and their management made up of the vegetal, the technological, the communal, and the indigenous.[71] Tavares's work emerges from a critical archaeological perspective on the conflation of extracting industries and monitoring technologies in the Amazon rainforest and develops an ongoing argument and practice concerning nonhuman rights and the space of the urban forest.[72] Both projects aim to ground space for designing other assemblages of sensing techniques and relations with the environment, including "a multiplicity of subjects and inhabitations," to quote Gabrys.[73] We return to this topic at the end of the book (chapter 7).

THE BOOK CHAPTERS: A MULTISCALAR APPROACH

Our focus is decisively multiscalar. Chlorophyll, plants, canopies, forests, prairies, and global objects such as soil or the biosphere are examined in the book through their entanglement with glass cases, photographic paper, measurements of light, datasets of images, and image-based models of growth. Through a series of cases, we focus on the question of the surface as a recursive interplay of vegetal and media practices, from early experiments and observation of vegetal matter and photosynthesis to plant physiology since the nineteenth century. We frame much of our discussion by recent debates in remote sensing, machine vision, and AI techniques of calculation of agricultural and other landscape surfaces.

The first chapter picks up on the material culture of glass that has become essential for understanding the scale and chemistry of plant life. The reactivity of the surface of plants to rays of light suggested the usefulness of transparent sealed vessel, which were essential for Joseph Priestley's late-eighteenth-century experimental demonstration of the ability of plants to generate their self-preserving atmosphere. The chapter traces how this material culture of containment became, in Gabrielle Hecht's terms, an interscalar vehicle as it traveled to the planetary scale as well as to the microscopic vegetal cell.[74] That is, it grounded, on the one hand, a logistical network of greenhouses connected by Wardian cases as mobile architectures, which sustained the circulation of living plants linked to the expansion of plantations and colonial trade. On the other hand, the entrance of plants in the interior space of glass globes in the laboratory also became a model for the endosymbiotic passage of chlorophyll to the interior of the transparent walls of the vegetal cell. Very different scales of the interweaving of plants to light were affected by the emergence of the technique of the transparent envelope.

The second chapter is focused on a single scale of the plant and its leaves. It deals with how the parallels between the photosensitive surfaces of plants and photographic materials were recognized in the nineteenth century in the context of plant physiology devoted to the study of vegetal growth. The convergence of these two different forms of inscription of light is discussed as a space of transfers between vegetal surfaces and imaging practices, where cameraless techniques of photographic measurement as well as other media, such as time-lapse photography and other early cinematographic techniques, rendered plant matter as active and sentient. Featuring the work of plant physiologists such as Julius Wiesner and Wilhelm Pfeffer, as well as early experiments with moving images by pioneers such as Henderina V. Scott, the chapter discusses how the surface of leaves was explicitly related to photosensitive media in a series of experiments where isolated vegetal matter started to be understood as living registers of light.

Realizing that plants play a key role in the maintenance of the atmosphere has crucial implications for the multiscalarity of vegetal phenomena. Vegetal life is not merely tied to plant physiology but is part of planetary processes. In 1903 Kliment A. Timiriazev called this the "cosmical function of the plant."[75] In chapter 3, we trace how plant physiology (as well as other domains of scientific knowledge such as thermodynamics and soil science) articulated a movement through scales that characterized the background of the first models of planetary life, developed by the biogeochemist Vladimir I. Vernadsky and the mathematician Alfred Lotka, among others. We focus in particular on Vernadsky's biogeochemical notion of the biosphere from the 1920s, not least because it has resurfaced in the past years in the context of discussions about the Anthropocene and Gaia. Vernadsky's biosphere is presented as a take on the physiology of the earth's uppermost crust, a surface turned into an interface layer around the planet. In this, the planet is rendered biochemically sensitive as a living film modeled after the surface of plants. The term "film" is not metaphoric: it describes the layers and surfaces identified in scientific terms around the early decades of the twentieth century, a period also defined by the intensive influx of new chemical materials in the context of imperial geopolitics.[76]

Parallel to the theoretical upscaling from the surface of the plant to the biosphere, two related technologies accelerated the industrialization of agriculture: the production of synthetic ammonia and aerial photography. In chapter 4, we discuss how large-scale agricultural programs were practiced as the circulation of images resulting from these chemical and mechanical techniques. While we focus on the agrarian Spanish "inner colonization," where the total surface of irrigated lands was doubled, we analyze the case regarding the relationship between different large-scale agriculture and land settlement programs of the twentieth century. We show how these programs were understood as processes of media and technology, where land was managed through visual operations and monitored as such. We emphasize the role of image-based technologies and relate them to contemporary practices such as precision farming, where agriculture has evolved to become a practice of the image extended by mechanical devices operating in the air and on the ground.

The operationalization of the uppermost crust of the earth as a surface managed through cultural techniques of images and calculation goes beyond agriculture. Thus, in chapter 5, we track how analytical knowledge about the surfaces of the world—landscapes and territories—shifts to synthetic knowledge about the surface of images. We analyze the shift where the notion of ground truth is no longer specific to the surface of the ground as a geological or geographic reference point, even if it plays a central role in calibrating aerial and satellite remote sensing for mining and other extractive industries. Instead, ground truth becomes read through the constantly evolving set of relations among environments of images increasingly populated by the complex devices, infrastructures, and protocols in Earth Observation systems. In addition, we build an argument about the synthetic landscapes experimented with within current contexts of AI, which we will refer to as "fake geographies," using a term already proposed in computer science. These questions deal with what becomes decipherable as an image and as a landscape in systems that function primarily through data as their input.

Chapter 6 continues the scalar trip from the microscopic to the single plant, from the agricultural plot seen from the air to the planetary, and from the scale of vast prairies such as in the American Midwest in the early twentieth century. Grasslands gave rise to a connection between imaging and modeling surfaces that preempts some forms of more recent environmental observation systems. In particular, we examine the work of botanist Frederic E. Clements, whose quantitative plant survey methodology relied on images, statistics, and techniques of counting. His statistical approach to the spatial characteristics of plant communities is tightly linked to the systematic elaboration of image-based indicators of surface characteristics, such as the photobotany practiced by Cold War intelligence programs. We build on the research by Robert Gerard Pietrusko on techniques of indication and how ecological knowledge was integrated into remote sensing operations, such as the ones found in current programs like the Advanced Plant Technologies project at the Defense Advanced Research Project Agency (DARPA) in the United States. The knowledge practices gradually developed from the late nineteenth century to the first decades of the twentieth century are but one reference point in establishing an argument that resonates far beyond the original domain: a statistical understanding of living surfaces is an insight into the broader discussions about media techniques of ecological control.

As we start the book with the images and logistics of uprooted trees, we finish with the forest itself. In the last chapter, we extend our discussion into transparency and opacity as they appear in the interior spaces of forests. The chapter focuses on the consequences of images entering and emerging from the forest volumes. We describe the concept of light climate, so named during the second half of the twentieth century, and how it relies both on a technique for measuring landscapes of light through hemispherical photography and on a material culture related to weather warfare, including flash bombs and cloud-seeding operations. While forest canopies hinder operations of visual surveillance from above, architects and researchers such as Paulo Tavares and Hannah Meszaros-Martin have shown how visual techniques that go beyond optics have extended aerial monitoring below the tree canopies. The chapter focuses on the critical context of violence between transparency and opacity as they form elements of ecological aesthetics that emerge from scientific practices with plants and earth surfaces.

The last chapter also rounds up core themes of the whole book as far as they concern the environmental surface in relation to aesthetic tactics operative in ecologies of images, in human and nonhuman politics of space and bodies, and in the envelopes of finitude and emergence of vegetal life that express the inventive affordances of matter. As such, both the chapter and the book respond to recent discussions in media theory, arts, architecture, science and technology studies (STS), and speculative design: What forms and formats of ecology are being produced and imagined, and across what sort of planetary space? Where is aesthetics located, and where does sensing take place? This book suggests that we focus on the living surfaces that contract, register, and produce a multitude of multiscalar events.[77] Our opening images of floating trees and logistics of movement of biomatter are one series of images that articulate this multiscalarity in both allegorical and material ways: the rest of the book's chapters articulate a parallel track of surfaces of growth as an aesthetics of planetary surfaces.

1

FORESTS OF GLASS TUBES WILL EXTEND OVER THE PLAINS

An article in the September 1912 issue of the journal *Science* speculated about the applications that photochemistry might provide in the future. Its author was a scientist and senator of the Kingdom of Italy, Giacomo Ciamician, who based it on a talk he had presented at the International Congress of Applied Chemistry in New York earlier that year.[1] The main topic of his address was the possibility of harnessing the energy of the sun, as plants had already been doing for millions of years. Photochemistry, a discipline with barely a century of existence, was devoted to the study of the chemical actions of light on the surfaces of matter, and at the beginning of the twentieth century, photography was still the most salient of the multiple applications of its techniques and methods. While acknowledging the continued importance of research on the photosensitivity of surfaces, Ciamician's paper emphasized that the discipline could go further than any known uses. Other photochemical surfaces, he wrote, could be discovered and actively synthesized. Instead of light's transience being merely fixed, light's energy could be harvested in order to make available the power of the sun on an industrial scale, able to respond to growing consumer needs amid the industrializing, technological Western societies. Photochemistry could go *off the scale*.

If photochemistry had a specific scale, it was the molecular one. The processes it dealt with related to the transformations and exchanges of matter and energy at the level of ensembles of atoms surrounded by radiation. Ciamician's article imagined a different scale of impact, highlighting the need to address the effects that a photochemistry of the future could produce at the scale of the planet. He estimated, for instance, the amount of energy plants obtained from the sun and compared the result to coal reserves and their energy equivalent. His calculations showed that plants were much more efficient in converting light to energy than any known human-made idea, which made him postulate that a complete artificial photosynthetic process designed through photochemical means would lead to an outstanding source of energy once the scale of the ensemble of radiated surfaces of the planet was taken into account. Where plants already grow, specified Ciamician, they should be left to grow. In other areas where they did not, such as deserts, a new industry could transform solar radiation into valuable energy.

This scalar project was not free from political concerns. On the contrary, the colonial ideology in the background of this imagined convergence of technological progress and the planetary is explicit in some of the passing mentions found within the speculative plans. Large-scale surfaces would, after all, have to take into account existing territorial claims—or straightforwardly dismiss them: "The tropical countries would thus be conquered by civilization, which would in this manner return to its birthplace."[2] The photochemistry of the future that Ciamician supported relied on the scalar logic that characterized the realpolitik-styled imperial claims for territories across planetary surfaces. "The strongest nations rival each other in the conquest of the lands of the sun."[3] The manifesto-like text that prompted the need for research on photochemical processes was implicitly coupled to the logic of violence and dispossession that lingered in other forms and formats of captured energy.[4] The photochemical was infused in Ciamician's imaginary media landscape with a scale of a different order, that of the planet as conceived by the Western colonial project that had already incorporated the valorization and traffic of so-called exotic plants into its agricultural and economic repertoire while harnessing photosynthesis across the plantation systems.[5]

In this chapter, we examine a set of cultural techniques related to the research on photosynthesis and the metabolism and survival of plants through the overlapping scales that also characterize Ciamician's vision. We want to focus in particular on the relevance of a material aspect of this combination of refracting light and (glass-)encased worlds of plant life. Ciamician's story not only recalls the geopolitical contexts of energy but it also makes explicit the importance that glass had in the photochemical project in ways that would become a signature of the industries to come. While glass was already, and continued to be, itself a central symbol and material encasing modern industry, in this vision, it was also a means for harnessing photochemical energy: "On the arid lands there will spring up industrial colonies without smoke and without smokestacks; forests of glass tubes will extend over the plains and glass buildings will rise everywhere; inside of these will take place the photochemical processes that hitherto have been the guarded secret of the plants, but that will have been mastered by human industry which will know how to make them bear even more abundant fruit than nature, for nature is not in a hurry and mankind is."[6] If the British and European coal industries were characterized by the darkness, dirt, smoke, and stench of the cities, Ciamician's vision was a future of glass that moved beyond urban buildings and façades to entail glass tubes extending over the plains. This vision was not an original one. Glass as a bearer of light and future was present in the texts and imaginaries of his contemporaries[7] and, as we will see in this chapter, it featured prominently in the material culture of photochemistry: the chemical reaction of photosynthesis became understood enclosed inside plant cells, where the chloroplast stored sun's energy, while at the same time different artificial encasings of enveloped air—even a glass house—helped to store and transport plants across the planet. From the chloroplast to the cell to the different housings of plants, a multiscalar universe of environments of growth emerged. In this regard, we want to understand glass as what Gabrielle Hecht has named an interscalar vehicle and make use of her methodology to track a set of experiments, devices, and architectures where the presence of this material coincided with the overlapping scales we have described in Ciamician's case.[8] By "interscalar vehicle," Hecht refers to "empirical objects" that can connect "stories and scales usually kept apart."[9] This could mean, for example, "maps and photographs" (Hecht's examples), as well as glass vessels and other architectures that both are concrete material objects and can, at the same time, trigger leaps across different scales of reference and impact.

Since the early experiments in photochemistry in the eighteenth century, the observation of plant matter worked in and through a material culture of glass jars and other transparent containers. This material culture of containment became essential for understanding the scale and scope of plant life as well as for the research into chemical reactions such as those involving oxygen and air. Mini-atmospheres were theorized in several different national contexts of chemistry. Beginning with Joseph Priestley's late eighteenth-century experiments, the reactivity of the surface of plants to the influence of rays of light suggested the usefulness of transparent vessel, which were essential for the experimental demonstration of plants' ability to generate their self-preserving atmosphere. Besides laboratory research into plants and air, the experiments included something more directly practical and industrial too. For us, of special interest are these other instruments of transparency, such as the Wardian case or the greenhouses of botanical and agricultural stations. What they bring to the fore is that the different kinds and sizes of glassed encasings were connected to the (mostly colonial) circulation of plants around the planet. While we are interested in this logistical theme of plant movement across the planet, a certain uprooting with which we started our introduction, we also describe the role of such glass cases in the process of understanding plant respiration, which by the beginning of the twentieth century had come to be described as the assimilation of light by green chloroplasts enclosed in transparent cellular envelopes.

We are not building a historical argument per se, but we want to emphasize how the glass globe, from photochemistry to the circulation of plants, privileged an understanding of the interactions between light and matter mediated by means of the transparent envelope as an interscalar vehicle. If photochemistry came to contribute to the flooding of the world with photographs and their wonders, glass surfaces that enclosed the production of photosynthetic processes came to function as a parallel transformation.

DISCONTINUITY OF SPACE

Since early modern times, the experimental method in science has relied on glass in many ways. A glass case allows the observer to simultaneously isolate an object of study, control its environmental conditions, and observe its transformation. While the glass envelops the observed entity, the transparent surface allows observers to scrutinize the confined volumes inside. As such, glass cases embody an ambiguous relation to the discontinuity they create between their outside and inside. While they produce a separate volume of space and matter, the transparency of the walls creates an illusion of continuity with their surroundings. Light flows in and out, emphasizing how this glass containment is part of a visual culture of observation *and* visual culture in the most literal sense of practices of light.[10] Light ties the inside to the outside. The fish in the aquarium and the plant in the terrarium become objects of observation while they generate in the observer the sensation of being perceived at the same time. Even more, they also produce the uncomfortable feeling of inhabiting a bubble of a different scale. As glass globes and other transparent cases are hardly perceived as simple individual objects, they speak for entities of a different scale as well.[11] The entire planet, a cell, or the limits of the walled-in self: transparent globes, glass cases, and other similar containers share the ability to bring in different scales at the same time. They are templates for models of enclosed worlds and, as such, they have been part of the material culture of all sorts of experiments in the context of natural sciences.

We want to highlight a sequence of historical experiments that expand the interscalar character of the glass case. This helps us to arrive at the example that brought together plants, light, and the planetary and that fired, in turn, the developments that we address later: the bell jars of Joseph Priestley. But before arriving at this experiment remarkable for both chemistry and the plant sciences at the end of the eighteenth century, we want to go through a set of episodes where the use of glass involves not only a play on scales but fundamental questions of elemental media and their relation to architectural and cultural techniques that define insides and outsides. The examples here relate to a specifically modern episteme of experimental knowledge so as to underline the scope of our argument about plants, images, and surfaces.

The first of these is a seventeenth-century experiment by René Descartes, described in *Les météores*, which was published as an appendix to his *Discourse*. The experiment deals with the enlarged model of a raindrop he commissioned so as to better study the trajectories of light to understand in geometrical terms the form and structure of the rainbow. The story, in Friedrich Kittler's words, reads as follows: "Descartes asked a glassblower to create a simulacrum of a single raindrop one hundred times enlarged. This hollow glass globe was just the promise of a larger thought experiment, in the course of which the Cartesian point-subject approached the sphere from every imaginable angle. The subject itself thus acted as a ray of light coming from the sun through the raindrop and executing every imaginable reflection and refraction until the simplest sunlight finally disintegrated, according to trigonometric laws, into the spectrum of the rainbow."[12]

The three spheres mentioned by Kittler—the glass globe, the raindrop, and the large virtual sphere—are included in the diagram used by Descartes to explain the experiment (figure 1.1). On the one hand, a myriad of tiny dots surrounds the rainbow: they are the droplets whose effects on the path of the rays of light have been sketched through the glass model in the diagram. A giant sphere floating in the sky, the size of the thickness of a rainbow, represents the virtual globe in the thought experiment. It illustrates how the colored stripes in the rainbow result from the play of angles of the trajectories of different rays of light. This play of angles depicts the findings of Descartes with the help of the transparent sphere filled with water that he commissioned from the glassmaker: by moving one's head around the sphere, the paths of light could be tracked, and the reflections inside it and the refractions when light travels from air to water and vice versa could be observed and measured.

For Kittler, the experiment functions as a forerunner of the technique of ray tracing in computer graphics. In the domain of the digital image, ray tracing names a technique where the lighting of a virtual scene is calculated through the reconstruction of the paths of simulated rays that are recursively reflected by all the surfaces present in the space. This reference to computer graphics highlights an early intersection in this

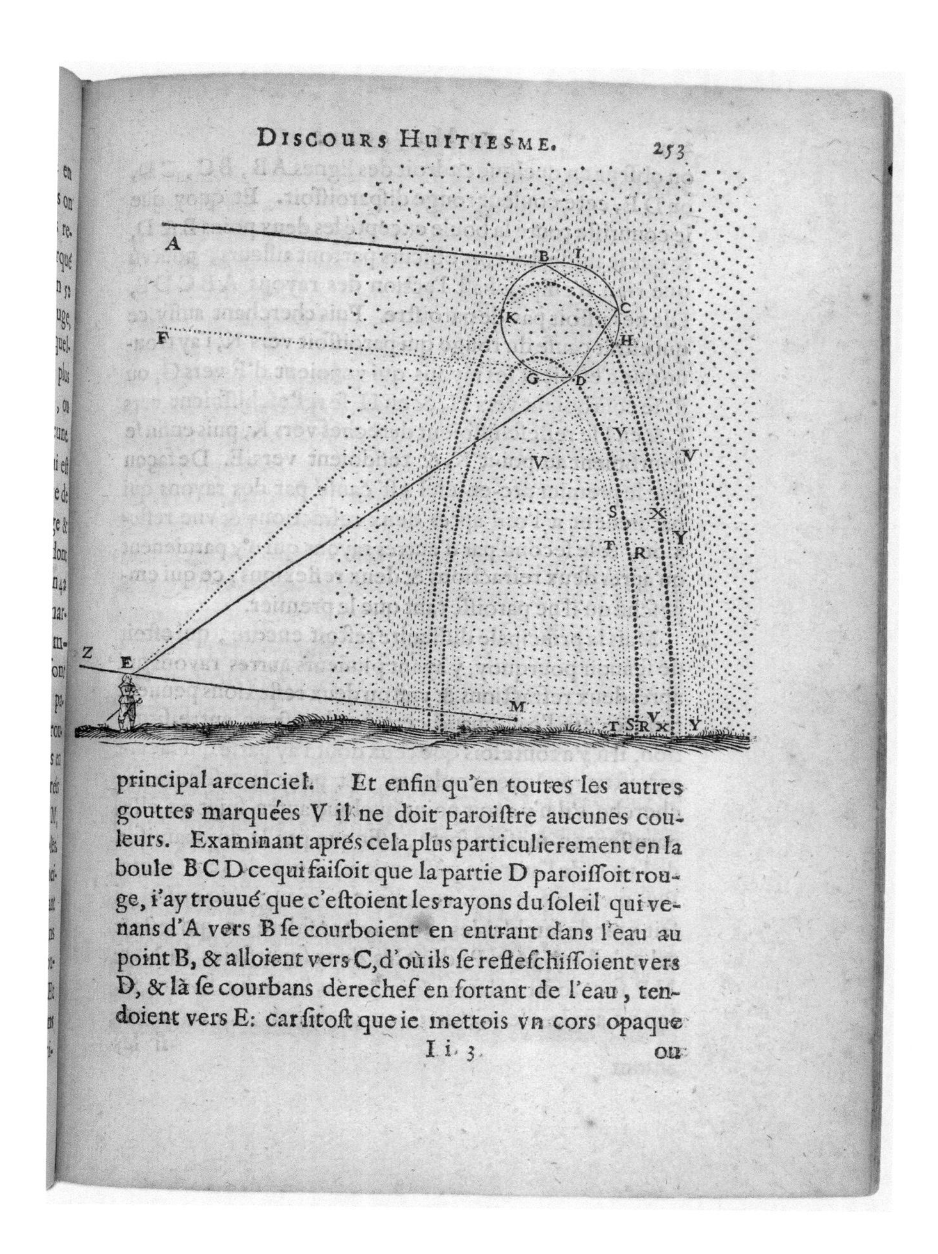
DISCOURS HUITIESME. 253

principal arcenciel. Et enfin qu'en toutes les autres
gouttes marquées V il ne doit paroiſtre aucunes cou-
leurs. Examinant aprés cela plus particulierement en la
boule BCD ce qui faiſoit que la partie D paroiſſoit rouge, i'ay trouué que c'eſtoient les rayons du ſoleil qui ve-
nans d'A vers B ſe courboient en entrant dans l'eau au
point B, & alloient vers C, d'où ils ſe refleſchiſſoient vers
D, & là ſe courbans derechef en ſortant de l'eau, ten-
doient vers E: car ſitoſt que ie mettois vn cors opaque

I i 3 ou

1.1

Illustration from René Descartes, *Illustrations de Discours de la méthode pour bien conduire sa raison et chercher la vérité dans les sciences, plus la Dioptrique, les Météores et la Géométrie qui sont des essais de cette méthode* (Leiden: 1637), 253. Source : Bibliothèque nationale de France ark:/12148/btv1b21000448.

interscalar diagram: a virtual enclosure which delineates image-objects through hidden lines.[13] Back to Descartes, the geometrical ray calculated through the sphere is both linked to the path of a visible ray of light, on the one hand, and to its representation as a line on the paper, on the other. The importance of lines as tools for thought as well as tools for drawing in the context of Descartes has been addressed specifically by Seán Cubitt, who argues that Descartes's *Dioptrique* represents a turning point in the relation between phenomena and the methods and metaphors used in the graphic descriptions of them.[14] As Cubitt shows, the adoption of lines as a legitimate representation of rays produced an unexpected entity when drawing light: the dotted line. Through it, the static immutability of medieval light—which did not travel but shone with the matter of space—met an "intermittence" that seemed "to go entirely against the concept of instantaneous transmission."[15] That is, while Descartes's writing emphasized the medieval belief in the unmeasurable stasis of light, the dotted lines he relied on in his diagrams pictured traveling rays, "splintering the continuous vector of time, and the expanding plenitude of light into the grain of surfaces."[16]

Now, while Cubitt's discussion leads to a specific genealogy of *visual* media, we are interested in understanding what role the glass sphere turned into a model, that of a raindrop, had in this period. While this occurred in Descartes's work at the level of the traced rays of light, the experiment with a transparent sphere brings to the fore such material divisions that operated in the same direction as the dotted line. The use of the experimental sphere was a legitimate tool for Descartes as it simulated a Euclidian model of the raindrops for his thought experiment. However, the hollow sphere not only represented a corpuscle of water but also addressed a genre of volumetric cuts in space that included the possibility of vacuum holes devoid of matter. Descartes supported the plenist theory that space is continuously filled with incompressible spheres of matter and argued for the physical impossibility of vacuum space. However, the long-standing debates over whether or not a vacuum could exist in nature relied on a number of experiments featuring sealed vessels, hollow glass, and glass balls.[17] Not by chance, Torricelli's experiment—widely discussed in France a few years after the publication of Descartes's texts—gave rise to a set of new experiments that led to the air pumps of Otto von Guericke and Robert Boyle.[18]

Otto von Guericke created with spheres the first device that extracted air, a mechanism that would be improved by Robert Boyle, who turned it into one of the most preeminent scientific devices, the *Machina Boyleana*. With this "engine," Boyle performed a series of well-known demonstrations of the production of a vacuum. He introduced candles into transparent containers to display how combustion would be made impossible if the air pump were applied. Some other experiments—cruel as they were—featured mice and birds who suffocated once a vacuum was produced. Shapin and Schaffer have famously discussed how this produced an experimental space inside the sphere with environmental parameters different from its outside; it also legitimated scientific experimentation as a practice separated from politics and religion.[19] But these experiments also certified an experience of volumetric discontinuity of matter (and life) in space. One could create artificial spaces that were devoid of air and hostile to life. When introduced inside the air pump, objects, processes, and living beings experienced the most abrupt disconnection from any material flow imaginable (at that time), as if thrown beyond the event horizon of a black hole. There is one crucial difference, though: light escaped from the interior, and objects abandoned to the horror of the vacuum could be observed from outside this spatial boundary, which did not merely detach insides and outsides but also opened aesthetic-epistemic questions of different kinds.[20]

An Atmosphere, Sweet and Wholesome

Fast-forward some 100 years: in 1771, Joseph Priestley would present his experiments on the composition of the gases that make up the air, creating an enclosed sphere. Certainly, his setup differed from air pumps. He did not remove the air but used glass vessels to alter the composition of the gases inside with the aid of a candle. With the burning candle inside, once the container was closed and sealed, he waited for the combustion to stop as the flame consumed the air. He did not build vacuum chambers but volumes devoid of oxygen (although it was only some years later that the gas would receive this more familiar name). As in the air pump, the absence of oxygen made living animals, such as mice, placed inside the jar next to the candle, die of asphyxia. The crucial difference in Priestley's iteration of the experiment was the introduction of a plant inside the vessel. Guaranteeing that no gas entered the jar—which contained the remaining "noxious air" after the combustion of the candle—he introduced a sprig of mint and observed how, in a matter of days, the plant "restored the air."[21] To wit, the plant was observed to regenerate the necessary gaseous composition for processes such as combustion and animal respiration to keep going. With a plant inside, the sealed space inside the jar "would neither extinguish a candle, nor was it at all inconvenient for a mouse."[22]

Priestley's experiment immediately became what we could now call an interscalar vehicle. The ability of a plant to modify its surroundings locally in such a way that other forms of life can benefit from it was perceived by Priestley and his contemporaries as a phenomenon that could be considered at the scale of the planet. In Priestley's words: "This observation led me to conclude, that plants, instead of affecting the air in the same manner with animal respiration, reverse the effects of breathing, and tend to keep the atmosphere sweet and wholesome."[23] Such artificial climates created de facto small-scale ecosystems, which seemed the exact inverse of the earlier experiments with vacuum cases: from taking life to restoring atmospheres and thus life. Put another way, even if arriving at the discovery by accident, Priestley took the bell jar as a valid model for the atmosphere as a whole, putting on firm ground the argument that plants exist in a mutual relationship with other forms of life by cleansing and purifying the atmosphere. "In this the fragrant rose and deadly nightshade cooperate,"[24] commented one of his contemporaries, acknowledging their accumulative effect. The enormous production of carbon dioxide by animal breathing and other sources could be countered by a different planetary agency. Understood in relation to their interaction with the atmosphere, the ensemble of plants, "the immense production of vegetables upon the face of the earth,"[25] could be considered responsible for the equilibrium in the atmosphere.

The increased knowledge about oxygen and a model for the atmosphere of the planet were not the only consequences of Priestley's experiment. Observed in relation to the use of other glass containers we have

1.2

Cover image of Joseph Priestley's *Experiments and observations on different kinds of air, and other branches of natural philosophy, connected with the subject . . . / being the former six volumes abridged and methodized, with many additions* (1786), displaying plants and mice inside glass jars. Public Domain. Source: Wellcome Collection.

addressed so far, the introduction of plants in the isolated space inside the glass chamber restored not only the material conditions for it to host chemical processes such as combustion or respiration but, in certain aspects, it also restored the continuity of space that the glass globes had broken. The ability of plants to process light and transform it into a chemical process on the surface of their body connected the interior of the glass bubbles with the space outside. In this sense, what was at stake was not just instruments and epistemic (architectural) environments in laboratories but how they defined constitutive boundaries, including those that are transparent.[26] In other words, the sensitivity of plants to light and the permeability of glass to its rays allowed an exchange between the inside and the outside.[27] The combination of the transparency of glass with plants sealed inside the isolated environments still enabled a connection to the outside—a connection established by rays of light and the chemical reactions thus triggered. This feature of plants—the ability to survive the isolation of a sealed space, the emptiness of oxygen inside the glass, and to even regenerate it—set the starting point of a circulation of artificial environments. Glass cases isolating plants and their generated atmospheres could be moved from one part of the world to another. The logistics of plants and light gave rise to the logistics of glass cases that transformed the environments of the planet under the rule of the colonial empires of the nineteenth century.

The Wardian Cases as Planetary Logistics

By the early nineteenth century, glass cases were not only hosting vegetal life and their "mini-atmospheres" but they were also enabling vegetal life to move across surfaces of the planet with new efficiency. Needless to say, plants had been on the move already centuries earlier aboard ships; this legacy was continued through new technologies and architectures of botany that both contained and mobilized living plants. The logistical apparatus of many European colonial states had created a set of enclosures through which samples sent back from other parts of the world were "purified" in different ways from their origins. As Londa Schiebinger summarizes,

> European naturalists thus tended to collect only specimens and specific facts about those specimens rather than worldviews, schemas of usage, or alternative

> ways of ordering and understanding the world. They stockpiled specimens in cabinets, put them behind glass in museums, and accumulated them in botanical gardens and herbaria. They collected the bounty of the natural world, but sent "narratively stripped" specimens into Europe to be classified by a Linnaeus or a Jussieu.[28]

Such enclosures were then both built (for sending back samples, whether seeds or live plants, and then for hosting them in an artificial climate of hothouses or museums) and written, referring to the logistical technique of classification as noted by Schiebinger and others who have investigated the connection between botany, colonialism, and the datafied ordering of the world. Many early naturalists, such as Hans Sloane, were "keen mover[s] of plants,"[29] establishing regular routes between London-based institutions and overseas territories such as Jamaica. Beyond a few expert naturalists, there was an increasing need to train and control how samples would be shipped back from different expeditions.[30]

Nathaniel Ward's 1830s glass-based innovation has become one of the most frequently cited examples to show both the emerging practices within botany of creating artificial climates and the legacy of transport of seeds and plants. It was meant to be easier to use by nonexperts, as it was thought not to need much maintenance. With the Wardian case, a significantly higher number of living plants survived trips across the oceans so as to be included either at Kew Gardens or in plant nurseries. Plants could be grown and studied in the artificial climate of the hothouses of London, for example, but also gradually shipped further into the colonies to establish new plantations (see figure 1.3). A planetary system of circuits and relays for plants was born that complemented the other architectural arrangements of colonial botany, such as herbaria like the one at Kew. As Zeynep Çelik Alexander has argued, such cabinet and spatial arrangements represented a form of a protodatabase too.[31]

The Wardian case was intended to replace earlier boxes and cases of transport by creating a sealed glass environment that protected plants against the outside. The risks of temperature variation, salty seawater, and lack of light were among the persistent challenges to successfully transporting live plants. Of course, many kinds of techniques had been in place—for example, the use of bog moss (peat moss) to protect any

transported plants or even the building of glasshouses on the ships[32]—but it was the Wardian case that has often been described as revolutionary for a variety of colonial and economic uses.

In addition, much of this early development of glass shifted from laboratories to the public and private spheres of life. In Britain, for example, glass cases and terraria started to become part of the Victorian indoors.[33] This shift culminated in the opening of the Crystal Palace for the Great Exhibition in 1851. Such new architectures enabled a spectacle

1.3

Wardian Cases used for shipping plants from Buitenzorg, West Java, in Indonesia. The image was taken at the Bogor Botanical Gardens that was founded by the Dutch East India Company in Indonesia. Photographer unknown. Collection Nationaal Museum van Wereldculturen Coll.nr. TM-10019035. Used with permission.

of transparency that disoriented the senses due to the new relation of incoming light. To historian Wolfgang Schivelbusch, this glass architecture was evanescent: "The uniform quality of the light and the absence of light-shadow contrasts disorientated perceptual faculties used to those contrasts."[34] As evanescent, such architectures seemed to shift the senses from a more focused observation in the glass instrument laboratories to a spectacle of shifting figures and grounds. Contemporary architectural commentators from the 1850s onward articulated such different effects as ethereality, light, and even the incorporeality of such spaces acting as new divisions of insides and outsides, which were sometimes also troublingly ambiguous. The architect—and later, director of the Berliner Bauakademie—Richard Lucae observed:

> As in a crystal, there no longer is any true interior or exterior. We have been separated from nature, but we hardly feel it. The barrier erected between us and the landscape is almost ethereal. If we imagine that air can be poured like a liquid, then it has, here, achieved a solid form, after the removal of the mold into which it was poured. We find ourselves within a cut-out segment of atmosphere.[35]

In many ways, we can say that the Wardian cases were earlier examples of such "cut-out segments of atmosphere" that built, directly or indirectly, on the scientific "lab" research into vegetal life and atmospheric gasses of earlier decades and also put the lab on the move beyond the parlor rooms or urban spectacles. Not that the urban was removed from the picture. Ward's observations of the purified air of such glass instruments took place amid the polluted London air that offered one context for the observation of local atmospheres, as bubbles of the sort where the plant itself became an instrument to read different locations of the city, an early biosensor, an indicator of sorts.[36] Besides the observations about the toxic city, Ward's version of the terrarium aimed to *create* atmospheric or climatic bubbles that shifted the material atmosphere and the imaginary of the location somewhere else. "Place the plant in one of my cases, where it has a constantly pure and humid atmosphere, and it will grow as well in the most smoky parts of London, as on the rocks at Killarney, or the laurel forests of Teneriffe."[37] Such scientists of fame as Faraday echoed Ward's prose, hinting at the possibilities of segments cut out of atmospheres to cater to animals and humans, part of emerging technological urbanism.[38]

With a twist to the story similar to Priestley's, the closed glass system became an observed unit of plant life and growth. Here, as demonstrated by Priestley and others, the glass unit also created its own conditions of living in a remarkable recursive loop of techniques often involving, implicitly or explicitly, photosynthesis. Cell processes, light energy, and atmospheres came together in such cases. In other words, a central element of fascination with glass was its relation to light, whether as part of the scientific laboratories or as new glass architectures, or indeed in creating temporary mobile ecosystems that facilitated plant movement from collection to the colonial centers of calculation such as at Kew Gardens in London.[39] Ward's work with enclosed spaces for plant life was thus related to this imaginary of creating *other spaces* inside London's murky storm clouds of industrialization (to echo John Ruskin), while also physically enabling further movement of plants.[40]

The glass cases actually moved spaces nested inside spaces, atmospheres inside *other* atmospheres, whole small worlds of conditions of plant life and living surfaces (see figure 1.4). Remember again the opening series of images of the book: vegetal life on the move, small-scale samples of living units being transported. As Luke Keogh writes, "The inside of a Wardian case was a micro-ecosystem of plants, soil, water, and captured sunlight, and with transpiration it functioned a little like our larger, outsider world."[41] This also meant, as Keogh notes, that not only were plants transported but also insects and plant diseases, inserting an interesting addition to what Michel Serres later termed the fundamental "parasite" of communication and transport.[42] The excluded third was not only the glass as a material that had to be fabricated, standardized, and mass-produced but also *the others* of those ecosystems of living surfaces, the unwanted guests that housed with the plants and soil. In a twist to this parasitical story, by the 1920s, Wardian cases had been modified to intentionally transport insects across the seas for experimental research, for example in Australia, although the transport of soil was in later decades brought into tight regulation so as to reduce the risk of unwanted (plant) diseases.[43]

1.4

"On the Imitation of the Natural Conditions of Plants in Closely Glazed Cases." Ward's book features illustrations of mostly indoors glass cases that mediated between the understanding of plant interiors as mechanisms of growth and the large-scale man-made glass architectures such as the famous 1851 Crystal Palace that botanist-architect Sir Joseph Paxton had designed. Image source: Nathaniel B. Ward, *On the Growth of Plants in Closely Glazed Cases* (London: John van Voorst, Paternoster Row, 1852 [first edition 1842]), 35.

FROM CONTROLLED ENVIRONMENTS TO THE MODEL OF THE VEGETAL CELL

The convergence of Priestley's experiments with the global logistics of the Wardian case and the glass architectures created a new environmental model for the private sphere of the bourgeois subject. Enlightenment light was able to flood in. Beyond any symbolism related to the interiority of subjectivity, glass was also a modern material that allowed questions of atmosphere and energy to be considered in a new light. It is in the context of such different environments of containment, cultivation, and growth that Peter Sloterdijk's philosophical quip about the development of Enlightenment into current atmotechnics makes the most sense to us: from the capture and modulation of sunlight to the production of air and atmospheres.[44]

Beyond the domestic and urban spaces, glass became, in the ways described above, a material connected to the scale of the planet. This was, of course, not a feature of glass material alone but was reliant on many forms of energy and extraction organized into large-scale networks since the emergence of Western colonialism and its multiple waves of logistical inventions. The labor of the enslaved population that made the plantation grow at unprecedented rates was essential to this system. In this book, however, we want to emphasize the role of plants that deployed their very particular form of respiration fueled by the energy of light when traveling inside the cases.[45]

1.5

Container for excised leaf used in experiments on photosynthesis and respiration. The petiole of the leaf is in a nutrient solution, the level of which can be adjusted. The air-stream enters at the top, leaves at the lower edge of the frame, and is dried over P_2O_5 before passing to the absorption tubes. Source: Herman A. Spoehr, *Photosynthesis* (New York: Chemical Catalog Company, 1926), 250.

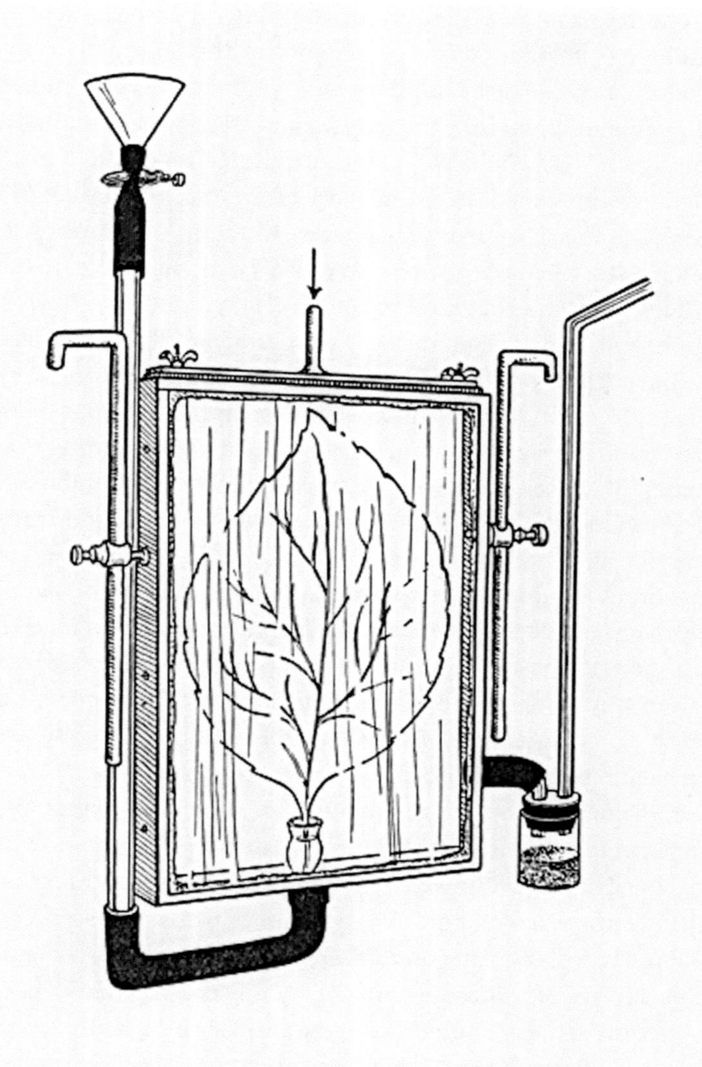

The reasonable reliability of the networks of standardized glass containers—in many ways, precedents of the shipping container that took over as the de facto standard and symbol of logistical mediation[46]—was matched by the inherent capacity of transport of biomatter: where encased animals would die, plants survived. The global logistical chain of living plants relied on the mystery of the "transmutation of light into bodies," as Russian plant physiologist Kliment A. Timiriazev had put it when he described photochemistry in his Croonian Lecture delivered to the Royal Society of London in 1903.[47] The reference to Newton's words, with their alchemic resonances, is not accidental.[48] As we discuss in more detail in chapter 3, at the beginning of the twentieth century, vegetal photosynthesis was understood in cosmical terms in the context of Russian science. Paraphrased by Timiriazev as "the real link uniting that glorious outburst of energy in our central star with all the manifold manifestation of life in our planet,"[49] the ability of plants to arrest and transform the energy of light had become an important experimental and theoretical question in the domains of plant physiology and photochemistry of the early twentieth century. Again, this question relied on a material culture of all sorts of glass vessels englobing vegetal matter (see figure 1.5) as the Russian physiologist summarized succinctly: "For 35 years I have been staring . . . at a green leaf in a glass tube, and breaking my head in vain endeavours to clear up the mystery of 'bottled sunshine.'"[50]

The question we are referring to is, Where do plants "bottle" and transform the rays of light? Where does their respiration take place? During the early years of modern chemistry, in the second half of the eighteenth century, a well-known experiment on this "mystery" began with plunging leaves of plants under water. Submerged, they cover themselves with a surface of minuscule bubbles of air. The mechanisms behind the production of these bubbles were a topic of scrutiny among practitioners interested in plants as well as in gaseous substances. In the 1750s, a series of systematic repetitions of the experiment led to the conclusion that the origin of the bubbles needed to be attributed to the surface of the plants themselves.[51] Decades later, after Priestley's research on air and the respiration of plants, it was proposed that the bubbles were dephlogisticated air—that is, oxygen—emitted through the surface of the leaves.[52] Later, it was observed that the bubbles were formed in only the green areas of the leaves, which led to the idea that the production of oxygen, their "principle of vegetation," was linked to the substance involved with that color.[53]

Driven by the tiny bubbles on the surface of the submerged leaves, the location of plant respiration was narrowed down to the scale of the pigments that made plants green. The fluctuating relation between leaves, their color, and their exposure to sunlight is already accounted for in texts devoted to them from antiquity, such as Theophrastus's *Enquiry into Plants* or Aristotle's *De Plantis*,[54] but the modern experimental observation of how these two initially unrelated phenomena coalesce—their pigmentation and their respiration through the absorption of the rays of light—helps to demonstrate the importance of glass, traditionally linked to optics, in the context of photochemistry and plant research. Indeed, plants continued to be enclosed in the interiors of glass bubbles after Priestley's experiment.

To understand better the relation between the respiration of the plant and the physics of color, a set of famous experiments was assayed that either placed plants inside tinted glass jars or that illuminated plants with single color areas of light refracted by a prism.[55] This set of experiments demonstrated the other side of the relation between plants and color: plants are unable to transform green light. That is, if placed under only green light, plants would perish. While they absorb most of the other colors of light, green is reflected. This is how pigments work; in the context of the research on the mechanism at play during respiration, the consequence of these experiments was understanding the fundamental role of the pigment on their surfaces. Their green pigmentation is a consequence—a leftover—of their respiration through their surfaces. The chemical interweaving with light—and not with the warmth of light nor other environmental conditions associated with sunlight—made plants radiate oxygen and reflect the color green as part of the same process. As a purloined letter (a reference to Edgar Allan Poe's story), their hardly observable breathing was, after all, their most salient visual characteristic. While *breathing color* could scientifically be a misleading phrase, this synesthetic interweaving of color, respiration, and atmospheres points out the singularity of the vegetal: plant surfaces became volumetric as respiration.[56]

The interplay of light, streams of air, and the enclosing of plants inside glass containers continued to be explored during the nineteenth century and at the beginning of the twentieth century. The experimental setups display how this narrowing down of plant respiration and light to the concept of photosynthesis and the molecule of chlorophyll occurred in the recurring observation of bubbles nested inside glass containers. Furthermore, with the aid of the microscope—also a device initially made from a recursion of spheres of glass[57]—vegetal cells were observed directly. And the microscopic observation of the cells finally made visible where the rays of nongreen light ended their cosmic travel.[58] Corpuscles of chlorophyll, later known as chloroplasts, populated the green tissues of the plant. They did not spread on an undifferentiated surface of the plant, however. They were part of the vegetal cells: they were placed inside the interior space of their transparent cellular membranes. The transparent container walling the cell trapped inside the green organelles, which held the substance that operated the transformation of light into matter. The relation of containment was utterly evident under the microscope, and by the end of the nineteenth century the idea was discussed that chloroplasts might have first arisen as independent organisms that were later trapped inside the transparent space of existing cells:

> If it can be conclusively confirmed that plastids [chloroplasts] do not arise de novo in egg cells, the relationship between plastids and the organisms within which they are contained would be somewhat reminiscent of a symbiosis. Green plants may in fact owe their origin to the unification of a colourless organism with one uniformly tinged with chlorophyll.[59]

Despite their differences, bubbles, cell membranes, glass spheres, and Wardian cases can be speculatively taken as similar objects—but of a different scale. They can be understood as different instances of an interscalar vehicle where phenomena observed in one of them, such as the glass sphere, resonate in the other. The chloroplast filled with chlorophyll, trapped inside the transparent membranes of the vegetal cell, echoes the plants trapped and transported inside glass cases, which echo at the same time the vegetal mantle oxygenating the atmosphere at the scale of the planet. In each of these cases, the transparent containers respond to the appealing visual characteristic of the breathing of plants: it is a breath that can be isolated whenever light enters it.

Is the scalar jump to the scale of the cell just a speculative hypothesis? Throughout his extensive work on glass, historian of science Kijan Espahangizi claims that the role of this colorless material in the context of chemistry and biology of the nineteenth and early twentieth centuries cannot be fully grasped unless its active character in the formulation of scientific concepts is taken into consideration.[60] In particular, he observes how initial descriptions of the transparent membrane of the cell during the formulation of the modern cell theory in the 1830s were affected by the fact that most of the experimental observations were taking place under a microscope where probes were placed inside flat glass vessels, the watch-glass. In his words, "the pioneers of this field of research Matthias Schleiden and Theodor Schwann associated the hardly visible cellular boundary with a 'watch-glass,'" thus making glass leap from the instrumental apparatus to an epistemic placeholder.[61]

Espahangizi's approach to enclosures of glass in and out of the laboratory recalls similar points that theorists of cultural techniques make. In his work, glass as a material is understood as an excluded third whose properties have contributed to shaping the objects studied through it. It does not merely stand in the middle as a dividing line of insides and outsides but creates a long row of subsequent practices and material effects from visibility to observation, chemical reactions to transportability. Glass itself was the center of the action, even if transparent, somewhat like the raindrop that acted as the quasi-invisible backdrop for tracing rays of light. As its use became more and more widespread, Espahangizi argues, the chemical interaction of the liquids with the surfaces of glass started to worry the scientific community, giving rise, on the one hand, to a conceptualization of chemical surface effects and, on the other hand, to multiple techniques of testing and standardizing glass.[62] The research on the surface effects of glass, in turn, gave rise to an epistemic object, the porous membrane. Significantly, Espahangizi remarks this porous membrane was going to be used, in the end, as a model for the walls of the cell itself. The cellular membrane was modeled on the porous properties detected on certain types of glass, to the point that even a physical glass cell was designed. It was an experiment where a glass vessel (the glass cell) was introduced inside a second glass beaker (the controlled environment). In recursive steps, the glass containers' nesting is performed simultaneously as the production of the isolation and the isolation of a model. In Espahangizi's words: "In a sense, the twofold history of glassware as a material boundary of experiment and as a medium of thinking boundaries converges in this experimental design: On the one hand Haber's glass cell incorporated the traditional as well as the new, modified laboratory glass types, on the other hand the whole setting was meant to be a model for the cellular membrane."[63]

As an interscalar vehicle, then, glass zoomed in even onto the scale of the cell. So far, we have shown how the ability of plants to transform the inside of a sealed glass sphere, rendered initially as a space devoid of life, gave rise to a continuous movement of cases with living plants that reshaped an important part of the living vegetal mantle on the earth. Glass containers, however, worked not only as the modular units of transport and containment in the colonial logistical chain: Priestley's experiment inspired other elaborations, which ended up shaping the idea of the transparent membrane of the cell with the green chloroplasts inside them. This double aspect of glass cases grounded global logistical chains that transformed landscapes and surfaces across the world (for example, terraforming through the plantation system) while they became a model for the elemental units of the organization of vegetal life.[64] In other words, the glass spheres moved plants and gave rise to a new view of plant matter. Moreover, this form of matter, as we will see in the next chapters, was deeply connected to the cultures of the image.

The Photochemistry of the Future

Beyond the laboratories of the nineteenth century, even beyond the logistical media of ships and Wardian cases, urban spaces of the twentieth became the locus for merging glass and images. Giuliana Bruno recalls such key theorists of early media of urbanism as Siegfried Kracauer, writing how the "shimmering light of the urban arcade" equaled the "flickering space of the movie theater," as though they expressed the same surface condition.[65] A fleeting and phantasmagoric disposition of light in motion rematerialized the world as a "surface tension" that "can turn both façade and framed picture into something resembling a screen."[66] Glass-covered cities spread, while glass and screens became, in many cases, almost interexchangeable.[67] This surface tension spread even outside the technological city as it was scaled up to landscapes seen from the air, like the visions of forests of glass tubes imagined by Giacomo Ciamician's photochemistry of the future. In a remarkable scene written by Paul Scheerbart, the whole skin of the planet is rendered as a twinkling fairy-scape of animated light: "The surface of the earth would change immensely if everywhere brick architecture were replaced by glass architecture. / It would be as if the earth had clothed itself in jewellery of diamonds and enamel."[68]

Janet Janzen highlights the prominent role that glass played in the science fiction stories by Scheerbart. Those featured, for instance, gardens that contained animated fragments of colored glass instead of plants, presenting a case of the wider aspiration of early twentieth-century colonial societies to transform nature and ultimately replace it.[69] However, the glass containers and architectures we have analyzed in this chapter show how this "improvement of nature,"[70] using Richard Drayton's words, had already been performed systematically through glass walls and glass cases as interscalar vehicles of light, plants, and atmospheric gases. While the key combination of glass and screen features prominently in the early twentieth century and the multimedia screen enclosures of the Cold War,[71] already hints of chemical and atmospheric elemental media had been featured in the experimental work and architectures of emerging modernity.

As we have emphasized in this chapter, glass responds to the astonishing singularity of plant respiration: the physiological locus of their breath coincides with the chemical locus of their green color. They breathe through light, connecting to a whole long circuit of possible techniques of tapping into, modulating, and recircuiting that light. Similarly, with the scientific realization of such capacities of plants, whole architectures have been built to experiment with light, temperature, and growth extending into the practices of light behind scientific agricultural experiments.[72] This chapter focuses on the circuits of connected cases and the nesting of recursive spheres. However, other circuits of light can be considered if we take into account the set of techniques that, from the nineteenth century onward, transformed the domain of the practices of light: photography. We will see in the next chapter how, as soon as the technique of registering light on photosensitive surfaces started to circulate, it was immediately assayed with photosynthetic surfaces, giving rise to loops of a different sort. For instance, in their *Essai de statique chimique* in 1842, Dumas and Boussingault already observe that the mysterious qualities of the green chloroplasts made them apparently "disappear" when photographed: "In fact, when we transport their image in the camera of Mr. Daguerre, these green parts are not reproduced, as if all the chemical rays, essential to the Daguerreian phenomena, had disappeared in the leaf, absorbed and retained by it."[73]

If the green vegetal matter was systemically enveloped in glass due to the optical characteristics of this material, the technical cultures of photography extended this interweaving to the domain of the image. This extension had already taken place if we consider that microscopy operates as a recursive zoom of glass lenses and projected images.[74] In this regard, photography gave rise to microphotography: the camera, instead of the eye, captured pictures of microscopic views that started to be systemically taken, reproduced, and circulated.[75] What's more, to obtain higher levels of visual amplification, the pictures of microscopic views were placed under the microscope and photographed again. A recursive loop of images gave rise to other images: images of plants, instead of plants themselves, were observed under the microscope.[76] This resulted in views of plants described as imaging apparatuses themselves, such as the case of Harold Wager plant photography—that is, photography *by* plants.[77]

"We have secured ourselves 'behind a barrier of perfectly engineered glass,'" writes art historian Kaja Silverman in a commentary of Andrei Tarkovsky's filmization of Stanislaw Lem's *Solaris*. This way, she continues, like the scientists in the movie, "we can 'study' an oceanic planet without getting 'wet.'" This oceanic planet, she goes on, is our world, "and it is through photography—rather than hallucinations—that it speaks to us."[78] We will confine glass to this chapter to plunge next into the photochemical oceans of photography with its vegetal hallucinations.

2

THE PHOTOGRAPHIC SURFACE OF THE PLANT

In the previous chapter, we showed how a long-lasting material culture of transparent cases characterized the observation and manipulation of plants in relation to their ability to respond and adapt to their environmental conditions. From Priestley's jars in the eighteenth century to early twentieth-century images of forests of glass tubes by Giacomo Ciamician,[1] from the Wardian case to the transparent membrane model of the chlorophyll cell, the enclosing of plants in transparent containers crystallized simultaneously into descriptions of vegetal matter and into a biotically reconfigured planet, the consequence of transatlantic movements and plantation cultures.[2] While vegetation had established itself as a phenomenon of significant impact on the actual habitability of the earth, this additional layer of transparent architecture reformulated it under the specific circumstances of a controlled surface of a globe. The glass case became the key cultural technique for a world shaped by logistics that simultaneously shaped the interior spaces of plants.

This colonial story of observation, extraction, and transportation came to feature alongside other logistical operations. During the eighteenth and nineteenth centuries, a historically situated rendering of vegetal matter had set the ground for collecting materials and performing experiments that featured the manipulation of plants. Since Aristotle's distinction between animals and plants as holders, respectively, of an *anima sensitiva* and an *anima nutritiva*, plants were perceived in the Western tradition as passive and insensitive beings, as creatures animated with a purposeless ability to react to their surrounding conditions.[3] During the first half of the nineteenth century, the mix of practices and treatises that would become the discipline of physiology still illustrated this distinction.[4] The two Aristotelian categories—*sensitiva* and *nutritiva*—pervaded in the mechanisms that were proposed to explain the movements of living systems, whether they were animal or vegetal. On the one hand, animals were acknowledged to behave with various degrees of autonomy. On the other hand, the movements of plants and the responses to their environmental conditions were attributed only to the action of external factors. Roots, for instance, were understood to grow downward because of the pull of gravity; the heat of the rays of light made plants orient themselves toward the sun; even the irritability of species such as the *Mimosa pudica* was explained as a physical reaction to changes in temperature, not very different from a hydraulic machine.[5]

As the nineteenth century advanced, this Aristotelian distinction was dismissed as many important changes took place in the life sciences. Charles Darwin's observations of the seemingly purposeful adaptation of plants to their environments, discussed in his widely read book *On the Origin of Species* (1859), suggested that they were able to perceive their surroundings, a proposal followed up by Darwin's contemporaries as well as his son Francis.[6] At the same time, in the German context, the life sciences being reshaped the laboratory-model of physics and chemistry, ready to test hypotheses such as plant sensitivity by experiments—which involved the building of research infrastructure, the training of specialists, and the systematic publication of experimental results.[7] Among these assemblages of experiments, a new set of graphical and visual instruments were adopted in many of the leading laboratories of the time. Self-recording instruments were of crucial importance and influenced by Hermann von Helmholtz's model of data gathering. They gave rise to protocols for measuring, describing, and communicating movements within the living. One particular graphical format, the Helmholtz curve, even became a method of analysis.[8] Prominent plant physiologists such as Julius von Sachs (and later, Wilhelm Pfeffer) produced a vast number of measurements out of the self-registration of temporal movements, which contributed to analyzing accurately the relation between stimuli and plant movements.[9] These instruments also allowed the trajectories followed by the plant roots and stems to be registered and visually recorded, setting up the basis for Darwin's proposal of the tip of the plant as a brain-like organ.[10] By the end of the nineteenth century, such work had resulted in the sensitivity of plants being accepted as a scientific assumption and becoming one of the most pressing research topics of the time.

In this context, photography, early moving images, and other visual media started to be used extensively in the plant sciences.[11] Photography in scientific laboratories had been in operation since the first daguerreotypes, integrated into the material cultures of the laboratory through its gradual connection to other instruments such as the telescope, the microscope, and the spectrometer.[12] As a consequence, in the mid-nineteenth century, photography played a key role in what historians of science Lorraine Daston and Peter Galison named mechanical objectivity, a configuration of epistemological convictions that made scientists consider working with self-registering instruments, or "photographers of phenomena," in the words of French physiologist Claude Bernard.[13] But the use of images went further than that. Beyond the laboratory, visual media populated textbooks and were used extensively for teaching purposes.[14] Moreover, public lectures sometimes showed photographic environments that were open and connected to living processes, such as the *Anordnungen* (dispositions or arangements) elaborated by German physiologist and pharmacologist Carl Jacobj, where organic, mechanical, and optical parts were assembled to transform in real-time the movements of the living into animated images projected on the walls of the research buildings.[15] In this regard, photography, cameraless practices, and early moving-image techniques expanded the milieu of the experimental culture around life sciences. The same experimental sites where plant sensitivity was described and scrutinized involved different types of surfaces—photosensitive papers, projection screens, and other instances of the image—reshaping also the milieus of knowledge; in the words of Austrian plant physiologist Raoul H. Francé, written in 1905: "Gone was the time of dead descriptions of leaves and blossoms. A new life had entered into botany."[16]

What was the new life alluded to by Francé? In the German-speaking context of the turn of the century, two debates about the notion of life coincided, from very different domains. First, in the circles of biological sciences, a neovitalist explanation of living mechanisms introduced by embryologist Hans Driesch, along the same lines as Henri Bergson's philosophic inquiries, was supported by several authors, including Francé. Broadly speaking, these critics of mechanical explanations of life revived a decades-old discussion between vitalists and reductionists about the existence of a substance or a principle grounding living beings, something that would be irreducible to the laws of physics and chemistry.[17] Second, a different debate concerned a distinction between the living and the inanimate. Moving images such as time-lapse sequences and other techniques in early cinematography mobilized a concept of life that film theorist Inga Pollmann has described as "cinematic vitalism,"[18] an umbrella term that refers to the vitality of the objects on the screen, to the animate movements in the machines that produced them, and to the vivid bond generated with the spectators. Not by chance, these images were initially addressed as "living images," a new epistemological unit that included at its core the media technological context of inscription.

These two parallel discussions raise a follow-up question: Did these debates converge into a cultural context of accepted analogies between vegetal life and moving images? Scholars in cultural and film studies, such as Janet Janzen, Teresa Castro, and Matthew Vollgraff, agree that while the space of the image amplified the otherwise imperceptible movements and reactions of the plant, the reappearance of a vitalist discourse infused early moving images with the animist character that several authors and filmmakers observed and further explored during the first decades of the twentieth century.[19] In this regard, while the parallel work on plant photography practiced and described by Harold Wager would suggest similar themes about "plant intelligence," he himself was mostly interested in the model of the capture of light by "plant lenses extracted from leaf epidermis."[20]

Such an important cultural moment of exchange between vegetal life and visual media can be traced in the technical and material cultures underlying the experimental practices of German-speaking plant physiology of the turn of the twentieth century. In the following discussion, we focus on a series of measurements taken in Vienna by one of the most distinguished plant physiologists of that time, Julius Wiesner. A close look at his experimental practices will show how vegetal and photographic surfaces were explicitly exchanged as part of his technique of surface photometry. For Wiesner, photometry was not only a technique to be used in practical domains, such as in photography or in the control of light environments, but also an elemental technique to be found in the vegetal realm. Plant perception, he proposed, was photometric, picking up on earlier research that had made it possible to isolate and compare intensities of light. Now though, the instrument of observation was the plant itself as an active agent of perception.

We want to elaborate this theme and argue that the two surfaces—vegetal and photographic—merged in a cultural technique whose continuous application characterizes contemporary image cultures: the ability to form surfaces out of measurements of light. This could be considered one part of what photography had become: a surface that captures light. But beyond this definition, as it was registered, the light became formative of all kinds of surfaces susceptible to being affected by it, including living surfaces. To support this claim, we point out how, in Wiesner's publications, photographic images and plant surfaces appear implicitly as ensembles of measurements of light. The traits of philosophical vitalism in his work, its relation as well as contrast to well-known supporters of mechanistic explanations of life—such as Pfeffer and Darwin—and the implications at the environmental scale of his elemental photometry are addressed later in the chapter.

Lichtgenuss, or the Appetite for Light

Toward the end of his career, plant physiologist Julius Wiesner pursued a decade-long research project concerning the interaction of plants with light that relied on an extensive gathering of photometric data.[21] Wiesner, one of the most prominent Austrian contributors to the German-speaking plant sciences community, was the director of the Research Institute of Plant Physiology at the University of Vienna and an influential figure in the well-known Viennese Biologische Versuchsanstalt, the Vivarium.[22] He stood out as an expert in technical microscopy and a recognized authority on applied physiology, with links to the paper industry in particular.[23] He became an authority in society and political circles, including the parliament, and, featuring the characteristic Viennese fin-de-siècle disposition to cultivate together the arts and the sciences, he was also a remarkable practitioner of natural photography.[24] A cyanotype attributed to him is part of the Metropolitan Museum of Art's collection (figure 2.1).

The concept of *Lichtgenuss*, which can be translated as "appetite for light" or "enjoyment of light," became Wiesner's main research problem during the last years of his career. His study on the topic, gathered in his 1907 book *Der Lichtgenuss der Pflanzen* (The appetite for light of plants), aimed to show "how the plant as a whole behaves in relation to the luminous intensity offered to it."[25] It was a project that sought to build a holistic approach to the plant in relation to light, instead of dealing with its specific microscopic or chemical dimensions, as had his previous works on the formation of vegetal matter and plant transpiration.[26] Wiesner relied instead on a careful practice of measurements of the light received by the surfaces of plants at different points of their body at different times of the year. He addressed each of these measurements as a value of the plant's appetite for light, being a magnitude that was characteristic of the adaptation of a species to a particular environment. Wiesner postulated two arguments: first, that each species of plant is characterized by a disposition to react to light, which he named its Lichtgenuss; and second, that this Lichtgenuss was quantitatively measurable, that is, that this disposition to react to light could be reduced to a series of numbers. Surface measurements, a third aspect in addition to light and plants, were formulated as a layer of comparative data about these living surfaces.

To carry on his research, Wiesner needed to be able to measure the intensity of light accurately or, in his words, "to proceed photometrically."[27] To do this, he redesigned a well-known photometer developed by chemists Robert W. Bunsen and Henry Roscoe, which was based on the sensitivity of photographic paper.[28] Wiesner's invention, named the Insolator, consisted of a hand-sized rectangular piece of wood onto which sheets of photographic paper could be placed under a black cover that shielded them from the action of light (figure 2.2). The experimenter could then pull out

2.1

Julius Wiesner, *Frustules of Diatoms*: microscopic cyanotype attributed to Julius Wiesner, displaying an arrangement of diatoms; ca. 1870, cyanotype, 9.8 × 7.9 cm. Metropolitan Museum, 2012.108. Steven Ames Gift, 2012. Public domain. Source: https://www.metmuseum.org/art/collection/search/301891.

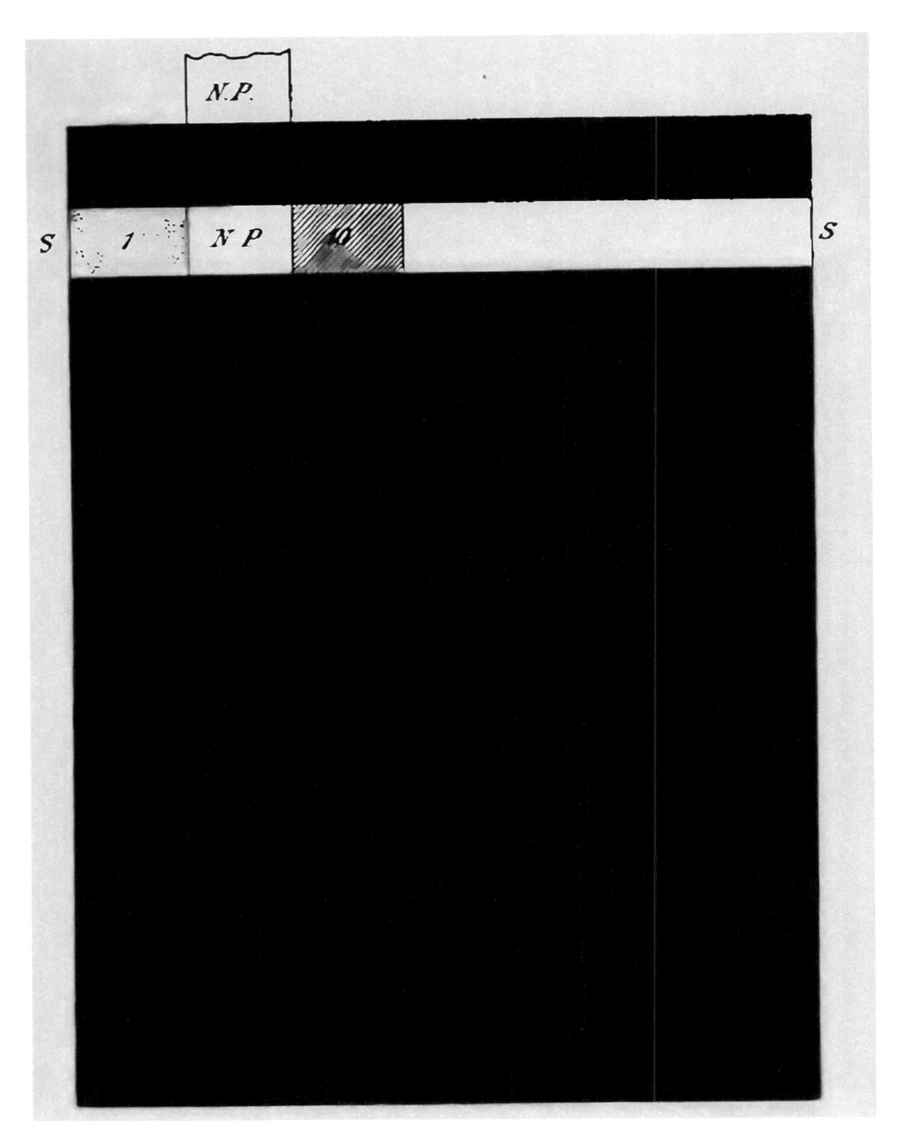

2.2

Illustration of the Insolator, a portable device used to measure the Lichtgenuss of plants; from Julius Wiesner, *Der Lichtgenuss der Pflanzen* (Leipzig: W. Engelmann, 1907), 15.

the paper and observe the blackening due to its photosensitivity. In order to easily compare this blackening to a reference value, a strip of paper with two so-called normal tones was attached to the border of the black cover. With this simple device, the intensity of light could be calculated with the aid of a chronograph by measuring the time needed for the photographic paper to reach the reference tone.[29] Additionally, there was a second, more elaborate way to use the Insolator with help from assistants. With this second method, he could place strips of photographic paper directly on top of the leaves of the trees and plants and thus measure the value of such *Lichtgenuss* surfaces that were harder to reach.[30] This striking overlap of leaves and photographic papers—with only a moment of explicit contact during the measurements—synthesized Wiesner's practice and ideas, as we see next.

The size of our modern-day smartphone, the Insolator was the portable photometer that Wiesner needed to measure the intensity of light anywhere and at any time. With this device, he obtained reference daylight curves and measurements of the Lichtgenuss of all kinds of Central European plants (figure 2.3). He even undertook several expeditions around the world to analyze the behavior of plants in extreme weather and light and height conditions, such as in the Arctic, the Sahara, the tropics, and the Alps.[31] Wiesner aimed to characterize each plant species in a certain geographical domain with a minimum and maximum value for its Lichtgenuss, taking into account the variations with the year's seasons. Plant geography, one important form of comparative planetary knowledge inscribed as maps and graphs since Alexander von Humboldt's expeditions, now included an additional layer of weather and light data. Plants became indicators of geographical variation, proxies of particular data, and even intelligence operations, as discussed in chapter 6.

Still, despite its apparent novelty, Wiesner's modification of the photometer invented by Bunsen and Roscoe was anything but original. In the context of photography, commercial devices that also relied on photographic paper, and which measured time and normal tones—such as actinometers and exposure meters—were already common.[32] The Insolator was really just a regular photographic exposure meter introduced and explained to the plant sciences community. The main function of the previous devices was the estimation of the amount of time needed for an image to form on the sensitive plate. Wiesner used them differently and in a

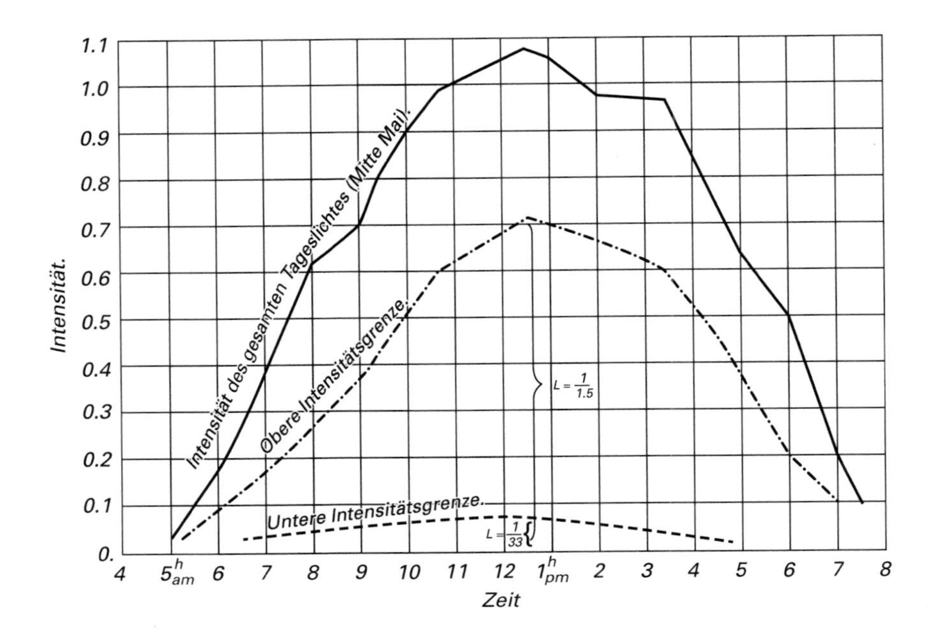

reversed way. This amounted to an alternative way of grasping the expanded sphere of operations of light that here came to define a scientific program of understanding vegetal growth as an image, that is, as a process that could be further controlled through this additional layer of photometric data.

REVERSE-ENGINEERING PLANT GROWTH

The notion of Lichtgenuss understood as an intrinsic feature of any plant was quantified by Wiesner in terms of a magnitude external to the plant: light measurement. To understand this procedure, the role of the Insolator and

2.3

Periodic measurements of the Lichtgenuss of a plant during a day by Wiesner; from Wiesner, *Der Lichtgenuss der Pflanzen*, 131.

its use as a description of the plant needs to be analyzed as a form and practice of inverted photography. Plant physiology started to be treated implicitly as a special case of what we would now call media, involving also an operational understanding of images. To see how these aspects of plants and (inverted) photography worked together, we can follow his routine step by step. First, Wiesner addressed a singular exemplar of a tree or a plant that was already formed and fully grown. He then measured with the Insolator the intensity of light where the tree or plant had grown, that is, where its living leaves were placed as part of the plant body in relation to its environment of light. Finally, he divided the obtained measurement by the intensity of the daylight background. The resulting number, the ratio between the intensity of the light at the surface of the plant and the background intensity of daylight, was defined as the Lichtgenuss of the plant.

The key idea lies in the following operation: if the measurement of light was done on any surface of the plant, this measurement instantly became the source for a value of its Lichtgenuss. In other words: in the measurements of the plant's appetite for light, the sun provided the light, the plant provided its spatial form—and no other vegetal parameter was involved. This mini-ecology of light was thus multiscalar in its operations, from cosmic rays to the very local site of registering the light in a spatial form. The multiscalar approach was made explicit in the words of Kliment Timiriazev, a Russian colleague of Wiesner: "certain rays of the sun, after having crossed without any modification immeasurable depths of space, on meeting on their way a chlorophyll granule cease to be light any longer, disappear in performing work."[33] This cosmic dimension of the logistics of light travel captures well the dimensions of living surfaces at play when considered as sites of inscription with planetary effects.

At first sight, this approach would seem to seek a parametric description of the plant's preferred surroundings as measured through variations of light intensity. To understand it only in descriptive terms, however, would be to miss its implicit aim of becoming a tool to explain the shape of plants. It is important to highlight that the plant physiologist was actively against any solely mechanistic explanations of phenomena of life.[34] Wiesner's subtle position argued at the same time for the sensitivity of plants when he wrote, "Shouldn't we speak of a sensation of the plant when one sees that it absorbs light, gravity and other external influences as stimulus?"[35] He, however, acknowledged the active, even Lamarckian, modeling effect of the environmental conditions, those of light in particular: "Light intervenes in the most varied manner in the plant's specific design process, either by direct action or by creating states which are hereditary."[36] His concept of the plant's Lichtgenuss linked these two agencies.

The concept of Lichtgenuss was not only a device to describe what was taking place, but it worked as a (teleologically tuned) explanation of the shape and growth of the plant. By measuring the intensity of the light on the surfaces of the grown-up plant, Wiesner attempted a reverse-engineering of the matter-forming encounter between the light and the plant. As the plant is supposed to be adapted through its shape to its environment, the intensity of the light in the point in three-dimensional space where the plant had placed a leaf ought to coincide with the average value of its Lichtgenuss. This assumed that the plant somehow had the ability to measure the intensities of light surrounding it and that it responded to these measurements by modifying its own shape. Wiesner assumed that plant growth was photometric and proposed, as a consequence, a photometric method, his Insolator, to quantify it.

This worked for Wiesner as an implicit principle, a latent assumption inherent in his concept of the Lichtgenuss: the measured light on the surface of the leaf coincides with the amount of light that the plant needs. In the logic underlying his method, the shape of a plant was the consequence of its photometric disposition and the ability to adapt to its surrounding conditions, which he designated as the *habitus* of the plant.[37] The plant's photometric adaptive ability was taken as a premise from which its shape was derived. The similarity of this line of thought to older vitalist principles in German-speaking science should not be ignored. As a matter of fact, the shape was, for the romantic *Naturphilosophie*, "the external impression in space and time of the organic formative forces,"[38] a principle which was defended by Wiesner's neovitalist colleague in Vienna, the already mentioned Raoul Francé, who assumed that plants expressed their passions through their movements.[39] Wiesner, however, did not propose a principle as such but devised a technique that embodied it. In his work, the polemics between vitalist and reductionist positions collapsed this way, inadvertently, in a photometric technique of the surface.

PLANTS AS PHOTOGRAPHS

So far, we have outlined how the principle underlying Wiesner's theory of the Lichtgenuss assumed that an elemental form of photometry was to be found inside plants. While the photometric measurement instruments had become a key part of deciphering light as an object of knowledge, this transposition assumed the plant as a form of registering surface, living but also "technical" in its own fashion. It is hard not to see in his work an explicit transfer of operations from photography to plant physiology: Wiesner's particular use of the Insolator links his concept of the appetite for light to that of the development of images on sensitive papers. As a form of reverse photography, instead of measuring the exposure time needed for a picture to be correctly formed, he measured the exposure time during which the surface of the leaf had been formed. In this regard, Wiesner's parallelism between the photographic formation of an image and the form of plants had already been perceived by a notable commenter of his time, Charles Darwin. In a letter to a Canadian fellow scientist about an earlier experiment by Wiesner, Darwin observed how the Austrian used terminology borrowed from photography to explain a series of movements in the plants.[40] In many ways, Wiesner's work was based on the claim, also made explicit at times, that the interaction of plants with light was similar to a photographic response. Beyond any pictures being formed, though, the analogy concerned operations of light and measurement.

It is worth observing that exchange between photography and research on plants was not uncommon at all in the nineteenth century. As historian of photography Luce Lebart has put it, the analogy between the effects of light on plants and that on photographic supports has been explored since the origins of photography.[41] Vegetal leaves, the skin of fruits, and juices extracted from plants had become a source—as photosensitive surfaces—for experimental practices of photography, such as Mary Sommerville's practices with vegetal juices, John Herschel's anthotypes, and Kliment Timiriazev's vegetal spectrum prints.[42] Observed against the background of these experiments showing connections between photographic and vegetal photosensitivity, what distinguishes Wiesner's work is that he reversed the direction of the relation: rather than exploring vegetal matter as photographic support, he looked instead for a photographic model of the vegetal morphogenetic drive. As argued above, this model was founded on the elemental technique of photometry, which already showed a data-driven approach to this combination of vegetal and photographic surfaces. In his book about the appetite for light in plants, he even proposed a classification of plant leaves based on their affinity to the exposure to light: leaves could be *photometric*, *aphotometric*, *euphotometric*, or *oligophotometric*.[43] In some sense, the claim was that the leaves developed light-producing—in photographic terms—vegetal matter as a result. This was a powerful idea that also resonated with different practices trying to make sense of the energetics of light as it impacts living surfaces: Wager's plant photography would be one of such practices where the leaf becomes the site of transformation of solar light into photography, or at least, where "the chemical action of light induces patterns in the shape and distribution of the chlorophyll,"[44] But in another register, the leaves of plants were perceived as the *factories* of organic matter. In Timiriazev's words: "Nature does not possess any other laboratory for the formation of organic matter, except the leaf, or, more strictly, the chloroplast. In every other organ and organism, organic matter is merely transformed; only here does it arise anew from inorganic matter."[45]

Wiesner's experimental practice shows how photography brought to the plant sciences and other environmental disciplines a new cultural technique, surface photometry. He understood plants as sensitive beings sculpted by light. They did not grow merely as accumulations of vegetal matter reacting mechanically to external forces, however, nor as free organisms with the ability to experiment with their own form. From his point of view, they were molded by light as if they were developed photographs of sorts. In this regard, the Insolator allowed him to collect Lichtgenuss tables with minimum and maximum values of different species at different times of the day and year, not very different from the exposure tables published by chemists and photographic manufacturers (figure 2.4).[46] With such aid, one could roughly predict the adaptive reaction of a plant if its surrounding light conditions changed: it would grow and evolve until the exposure times corresponding to its species, its Lichtgenuss, could be met. Lichtgenuss, or the measurement of the intensity of light where the plant had formed, was the measurement of the exposure time that kept the form of the plant alive—a living photograph

Lichtgenuß einiger mitteleuropäischer Bäume und Sträucher [1]).

(Nach in Wien und Umgebung angestellten Beobachtungen.)

	L (min)	Intensitäts-maximum	
Buxus sempervirens	$\frac{1}{108}$? [2])	0.012	Freistehender Garten-strauch.
Fagus silvatica	$\frac{1}{85}$? [2])	0.015	Freistehender Baum, Gartenform.
Aesculus hippocastanum . .	$\frac{1}{83}$? [2])	0.015	Freistehender Baum, Gartenform.
Fagus silvatica	$\frac{1}{60}$ [3])	0.021	Geschlossener Bestand.
Aesculus hippocastanum . .	$\frac{1}{37}$	0.023	,, ,,
Carpinus Betulus	$\frac{1}{36}$	0.023	,, ,,
Acer platanoides	$\frac{1}{35}$	0.023	,, ,,
Acer campestre	$\frac{1}{43}$	0.030	Freistehender Baum.
A. Negundo	$\frac{1}{28}$	0.046	Geschlossene Baum-gruppe.
Quercus pedunculata . . .	$\frac{1}{26}$	0.050	,, ,,
Ailanthus glandulosa . . .	$\frac{1}{22}$	0.063	Freistehender Baum.
Thuja occidentalis	$\frac{1}{20}$	0.070	,, ,,
Populus alba	$\frac{1}{13}$	0.086	,, ,,
P. nigra	$\frac{1}{11}$	0.118	,, ,,
Pinus Laricio	$\frac{1}{11}$	0.118	Kleiner, nicht dichter Bestand.
Betula verrucosa	$\frac{1}{9}$	0.144	Üppig entwickelter Gartenbaum.
Liriodendron tulipifera . .	$\frac{1}{7.5}$	0.186	Einzeln stehender Gartenbaum.

2.4

A vegetal exposure table, showing minimum and maximum values of Lichtgenuss for some Central European trees and shrubs, measured by Wiesner; from Wiesner, *Der Lichtgenuss der Pflanzen*, 153.

of its own species as well as a datafied way of optimizing particular kinds of growth conditions. We will come back to this conceptualization of vegetal growth later, as it is where the early formulations of living planetary surfaces already incorporated a hint of a model of planetary computation.

SURFACE PHOTOMETRY AS A CULTURAL TECHNIQUE

What was taking place in practices of plant physiology had its counterpart early on in photography. Treating the shape of plants as if they had photographic character was one indication of a broader context of how the light was being handled and understood through its effects. Plants' shapes emerged as the result of the encounter between imperceptible variations of luminous intensity and the plant's photometric need for light, which echoed French physicist François Arago's remarks about photochemical surfaces. For Arago, photographic images emerged as the encounter of both the "action of light" and a "sensing substance."[47] This suggests that Wiesner's work can be read from the perspective of an alternative tradition in early natural photography that cultural historian John Tresch has described in relation to the experiments carried out by Arago: the interest resided not in "the objects depicted on the silver plate, but rather in what the process and rate of its development revealed," that is, the registration of "invisible phenomena unfolding over time."[48] This shift from the pictorial to operations of measurement, registration, and comparison was a fundamental step that also, for our book, features as a central recurring theme.[49]

In this regard, Wiesner's photometric-driven morphology embodies elements of this early photographic constellation, which leads us to a point argued by art historian Vered Maimon: early practices of photographic imaging were related to an underlying focus on motion and temporality rather than on "substances and matter."[50] In Maimon's words, they were not meant to represent a static order of things but a register of "an evolving visual map in which vital forces mark themselves as they unfold in time."[51] This is precisely the vitalist principle we have excavated in Wiesner's concept of Lichtgenuss as a cultural technique of light and surfaces.

To think of Wiesner's methodology in photographic terms brings us back to the position where we started. Plants, photometry, and photographic images are trapped in a codefining loop in Wiesner's work: from such a vitalist perspective, the shape of a plant emerges as the expression of an invisible principle, its Lichtgenuss. But Lichtgenuss also implies elemental photometry, a technique that emerges as photographs, which become the expression of vital principles again. Wiesner mapped the relation between the surfaces of the plant and the unknown forces driving its growth into the relation that photographic surfaces had with photometric acts. A key idea is that this relation is recursive. The vitalist accounts of visual media mentioned earlier are examples of this, and others follow in the final section of this chapter. But for the moment, we want to focus especially on the theme of recursion.

To understand Wiesner's surface photometry as a cultural technique means to understand it through its self-referentiality, "a pragmatics of recursion."[52] This involves, in Bernhard Siegert's words, situating it before the grand epistemic distinctions it is involved in.[53] In Wiesner's case, this distinction concerns the "natural" versus "artificial" photosensitivity, or reducible versus nonreducible vitalism. The technique does not precede these distinctions, but it reshapes them into complementary configurations. There would be no cinematic vitalism linked to films featuring plants in movement without the operations of a scientific culture of measuring plants through images. Surface photometry as a cultural technique emerged as a transfer of the form observed in plants onto the medium of photometric photography—or in broader terms, photography as measurement.[54] And as such, it added photometric characteristics to plants as well as it added life-like singularities to photometric surfaces.

In addition, there is an aspect that needs to be emphasized concerning Wiesner's role in understanding surface photometry as a cultural technique. His work on Lichtgenuss brings in a clear example of the spread of photometric photography and the changes it operated in interweaving images and environments. However, this was not entirely new, as similar photometric transfer operations were taking place in other techno-scientific domains. The interest in his work relies not on any particular foundational claim or even on its long-term impact, as despite Wiesner's systematic and careful work being lauded by his contemporaries, his notion of Lichtgenuss and his later observations on the effects of plants barely outlived him in scientific contexts. Instead, the value lies in how these techniques are milestones in the historical and materially contingent hinge between imaging techniques and living beings, which pervaded the decades around the turn of nineteenth and twentieth centuries and which was later, as discussed in the next chapter, scaled up to the planetary surface.

SURFACE PHOTOMETRY ON AN ENVIRONMENTAL SCALE

Together with the color green,[55] surfaces are one of the most salient features of vegetal structures. "The plant's body is all skin,"[56] writes plant scholar Michael Marder. "It is for the sake of adhering as much as possible to the world that they develop a body that privileges surface to volume,"[57] continues philosopher Emanuele Coccia. Besides philosophy, such arguments can be traced back to scientific research. Wiesner's method of photometric measurements allowed him to address plants and trees as photometric surfaces (figure 2.5). In this regard, he proposed two additional technical surfaces in his book on the Lichtgenuss of plants in order to explore further contexts of applications of his ideas. These did not merely reproduce the photographic as a reference point but developed it in specific ways concerning concrete and abstract surfaces.

The first of the additional reference points was the *Lichtfläche* (the surface of light), defined as a surface that wraps the entire tree or plant.[58] In contrast to the paper sheets he laid on top of leaves, this surface was an abstract entity, akin to a mathematical concept, detached from any actual physical surface of the vegetal exemplar. The notion of Lichtfläche helped to deal with objects on a bigger scale, such as the crowns of trees or the extensions of grass. It was an epistemic tool that allowed Wiesner to arrive at a surface entity defined only in relation to its interaction with light. That is, it allowed him to describe a shrub, ivy, or rainforest palm. It was

2.5

Julius Wiesner, photograph of a tree crown that has adapted to the lighting conditions due to its position next to a building; from Wiesner, *Der Lichtgenuss der Pflanzen*, 75.

a mathematical coating,[59] to use the words of Georges Bataille, reducing complex assemblages of vegetal matter to surface shapes. Significantly, no reference to plant matter was even present in the name of the concept: surfaces of light.

Wiesner proposed the concept of Lichtfläche in order to calculate the green density of canopies. That is, he considered canopies as a geometrical area that could be compared with the sum of the individual areas of the leaves in the plant. Building on this scalar assumption, the ratio of these two extensions allowed him to quantify a measurement of the density of the foliage and its porosity in relation to light. This is a significant contribution as it is directly linked with the current notion of the leaf area index (LAI) that is calculated from satellite images and widely used in environmental research.[60] The utility of this definition had, in fact, already been detected in the 1920s by Russian biogeochemist Vladimir I. Vernadsky, who used it to calculate the density of what he called living matter in vegetal formations such as forests or prairies in order to compare them to the density of agricultural fields, which turned out to be of a lower value.[61] We will return to Vernadsky in the next chapter.

In addition to the Lichtfläche, Wiesner found another technical surface in a standard color chart available at the time, Otto Radde's so-called Farbenskala. The Farbenskala had already become an established means of color communication in scientific and trade applications.[62] This color scale allowed Wiesner to identify and classify the different greens observed in plants. By physically holding the metric scale on top of a vegetal leaf, Wiesner measured and compared the different shades of green present in plants and trees. This procedure made the plant leaf into a standardized image-like object that was codified with the tool that the industry used to regulate the correct circulation of colored commodities, furthermore integrating the observation of nature into the context of infrastructure of technical standards. This became a way of measuring the fluctuations of the green of leaves, their relation to the amount of chlorophyll, the effects of different lighting conditions, and many other phenomena. Notably, this colorimetric approach to the plant allowed him to perceive otherwise indiscernible variations that became understood only through the measurement.[63] Technical standards were to help in mapping thresholds beyond the observational capacities of the unaided human eye.

Images thus opened up new possibilities for measuring, while the ontological boundaries of what an image was were broadened in the process. Color values, intensities of light, and the flattening of the surface made up this experimental toolkit. The toolkit was, however, only one example in a row of experimental instruments and photographic environments that were transforming the discipline into a controlled laboratory practice.

FROM SURFACE PHOTOMETRY TO THE CONTROL OF ENVIRONMENTS

"The 30th of March, 1893, at 10h45m I observed in the Viennese Augarten a total daylight intensity=0.427."[64] Wiesner's diary report describes one of his acts of measurement. From a formal point of view, it constitutes just part of the gray literature of scientific knowledge operations: noting down observations as data.[65] The reader encounters a cluster of precise data in the text as one would imagine Wiesner's execution of the measurement itself: an interruption in the promenade, a sudden "explosion in the garden,"[66] recalling Carl Schorske's terms, performed perhaps with the engineered movements of Ulrich, Robert Musil's famous character.[67] As a scientific practice, it also responds to what historian of science Deborah R. Coen has described as the characteristic manners of a late nineteenth-century Austrian experimentalist: "a naturalist who haphazardly collected and classified 'isolated facts'"[68] outside the laboratory, in the open. This scientific persona is later described by Coen as "building hypothesis from the clues of accidental evidence."[69] The Viennese context, Coen explains, encapsulated a historically situated understanding of the practice of science: measurements in the open were the access to the expression of what was understood as nature's invisible forces, which surfaced as clues, signs, and evidence. These were indeed very peculiar kinds of signs, and they left a very particular trail of evidence, as it concerned light itself—technically visible light, but also what makes everything else visible and growing.

Such examples of the practice of Austrian plant physiology stood in contrast to the Prussian context, which was the more dominant model of that period. The latter was defined by the systematic and precision-led experimentalist being in close touch with industry. Hermann von Helmholtz's influence was obvious: Wiesner's Prussian colleague Wilhelm Pfeffer, the leading figure in German plant physiology, published the same year as Wiesner's book the details of a complex automatic regulator of illumination (figure 2.9) that, instead of measuring light, permitted it to be controlled accurately.[70] For the Prussian scientist, nature was seen as a poor experimenter, making it necessary to experiment systematically, with tight control of the environmental conditions.[71]

IMAGES AS A MEANS OF POWER AND CONTROL

Interestingly, Pfeffer's device for regulating light was part of a series of experimental setups which had led him to a well-known series of time-lapse photographs of vegetable growth; these were, in fact, the first time-lapse photographs to be produced.[72] Pfeffer's time-lapse images, produced between 1898 and 1900, were accelerated sequences of photographs that allowed researchers to easily see and recognize different types of plant movement, and they exemplify how a material culture of environmental control was also embodied in the technical space of Wiesner's surface photometry. As a media technique, time-lapse photography was devoted mainly to teaching purposes in the classroom, a cherished space for Pfeffer, who had introduced several other visual innovations. In the physiologist's words, time-lapses sought to provide students with a "correct and plastic image" of plant movement that was otherwise difficult to perceive.[73] Indeed, the behavior that the film registered had already been accounted for and measured a decade earlier thanks to a material culture of automatic self-registering instruments.[74] Now, the photographic was (again) directly part of the assemblage of observation and measurement. Pfeffer's time-lapses, however, exceeded the scientific domain, helping to popularize a view of plants as sentient and dynamic beings as we have already seen.[75]

In more detail, let us consider one of these time-lapses: the acceleration of growth of an exemplar of *Impatiens glandulifera* (figure 2.6). The sequence consists of one-minute long footage, compressing thirty-five hours of what is known as geotropic curving, that is, the tendency of a plant to

grow vertically against gravity. The initial scene shows the plant lying in a horizontal position, its rotated pot resting on a second, supporting pot. When the film starts, the tip of the stem begins an upward movement, gently curving the whole plant. The wandering tip oscillates, moving toward verticality until the stalk acquires, finally, an L-shaped form.

In the film, the plant seems to seek the vertical position itself instead of being pulled by light—or any other agent. In addition to its vitalist connotations, it is necessary to consider the actual physical sites where these sequences took place, namely, German life science laboratories and educational institutions. There, visual media were part of a program

2.6

Sequence of scene stills from Wilhelm Pfeffer's film on geotropic curving of the *Impatiens glandulifera*; from *Kinematographische Studien an Impatiens, Vicia, Tulipa, Mimosa und Desmodium von W. Pfeffer* (1898–1900), directed by Wilhelm Pfeffer.

based on "the gaze as a means of power to control the organisms,"[76] as historian of science and media Henning Schmidgen has remarked about the use of projection apparatuses mixed with living organisms in the context of German physiology. This control practiced by the gaze can also be observed in Pfeffer's time-lapse. In that sample of moving images, the plant in motion seems to be shaped neither by light nor gravity but by the framing of the photographs. The technical image pushes through as a force, an epistemically significant aesthetic in its own right. The images in the time-lapse recall one of the main ideas proposed by Joanna Zylinska regarding the relation between photography and life: "Photography is a formative practice of life not only because it represents our lives in various ways but also because it actually shapes life."[77] In the case of Pfeffer's time-lapse, the shaping of the plant by photography occurs in the eyes of the audience. The limits of the framing in the sequence constrain the image as well as the depicted growth. In the end, growth happens in the image and is shaped by it.

A MATERIAL CULTURE OF ENVIRONMENTAL CONTROL

In Pfeffer's work, the frame of the image is operationalized as an environment of control. Beyond a formal analysis of the images, an examination of the material culture surrounding the production of the time-lapses unveils a technical space around surface photometry where photographic processes are indistinguishable from other instruments of environmental control. In this regard, the specific case of Pfeffer's time-lapse of geotropic curving also connects to a very different experimental device, the clinostat,[78] designed by Pfeffer's mentor, Julius Sachs. It was improved and extensively used also by Pfeffer himself.[79] The clinostat is a mechanical instrument that allows experimenters to hold a pot horizontally so it can be rotated manually or automatically (figure 2.7). The apparatus aims to cancel the effects of gravity and light on the plant's growth. The slow rotation of the pot—one revolution per minute, controlled by a clockwork mechanism—makes the plant sense gravity and light sources evenly, with no privileged direction.[80] Consequently, the plant grows counterintuitively, along a horizontal line, parallel to the ground and the sky.

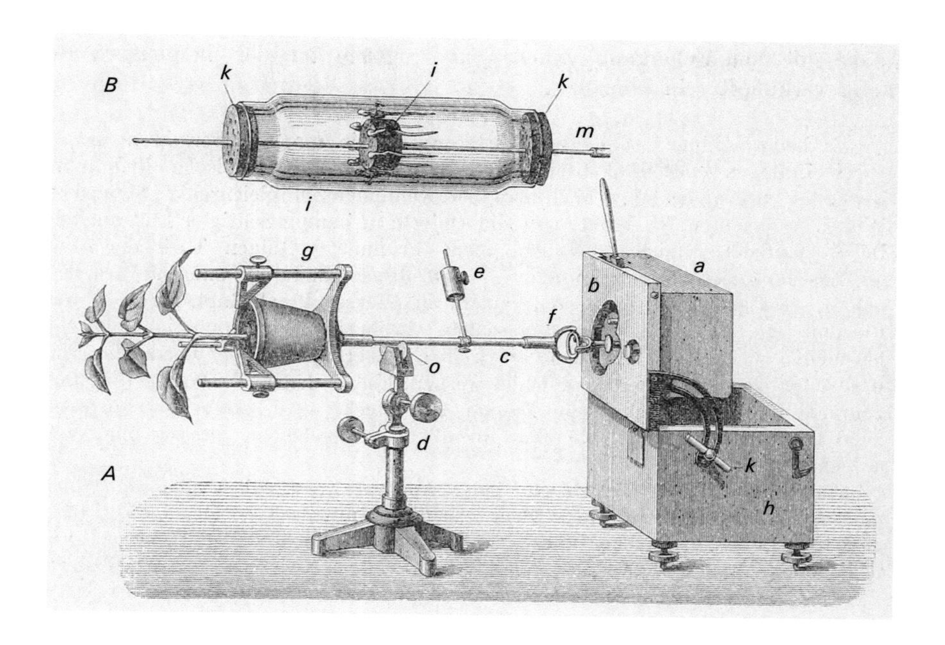

The clinostat relies on—and makes visible—the particular sensitivity of plants to their environment. "Their absence of movement is nothing but the reverse of their complete adhesion to what happens to them and their environment,"[81] writes Emanuele Coccia (although, in our view, there is plenty of movement that layers across different scales than animal movement). Plants are so deeply entwined with their surroundings that changes around them leave an impression on their shape; plants can be transformed by altering their milieus. This gravitation "canceling" apparatus, an example of historical *medium design*,[82] represents one case in the long repertoire of media techniques for controlling vegetal life. Understanding the rotating conditions of the plant as an altered environment that forces it to grow horizontally instead of vertically allows

2.7

Illustration of a clinostat; from Pfeffer, *Pflanzenphysiologie*, 570.

us to perceive this technique as tightly linked to Pfeffer's time-lapse. Here, such images are an operational part of the broader assemblage of visualization-manipulation.

To fully understand this point, we should observe that the German physiologist's sequences of images also rely on altered environments, similar to the ones involved with the clinostat. The thirty-five hours of observation were captured in a series of photographs to be used in an accelerated animation, requiring a careful synchronization of lighting conditions and rotational devices. As Kew Gardens botanist Henderina V. Scott remarked, when describing her process of building a cheaper version of Pfeffer's invention, uniform exposure and absolute rigidity of the apparatus were among "the very great practical difficulties" the setup needed to solve.[83] Although the experiments that led her to obtain the first time-lapse registered outdoors involved greater challenges than Pfeffer's—such as the varying light conditions of a greenhouse instead of indoor photography—most of the practical difficulties she documents also characterize the original setup. The production of the time-lapse was reliant on two key elements. First, it needed a mechanism to shoot a photograph and then to replace the sensitive film without moving the camera, so that the framing could be maintained as stable and useful. Pfeffer used a kinematograph that involved rotating a roll of celluloid film.[84] As this option was too expensive for Scott, she produced her photograph sequences with a kammatograph, where negatives were printed on a rotating glass disk instead of film (figure 2.8).[85] The similarities between the kammatograph and the clinostat stand out as if they were variations of the same hardware. The control of exposure is the light received by the photographic surface and the plant is another element that needs to be stabilized. For this purpose, the already mentioned mechanisms for regulating artificial lights were used for the night photographs, in particular: here, again, the clinostats were involved, this time explicitly (figure 2.9).[86]

Summing up, the same hardware that constrained the growth of a plant to a specific direction appeared within the mechanisms that accurately registered a sequence of plant movement on a sensitive surface. Moreover, the inclination prevented by the rotation of the clinostat was precisely the one registered in the sequence of images. It is as if the movements of the plant confined in the experiment with the clinostat had been

2.8

A rotating disk plate with the inflorescence of the *Sparrmannia africana*, registered by Henderina V. Scott in 1903; from Scott, "On the Movements of the Flowers," 780.

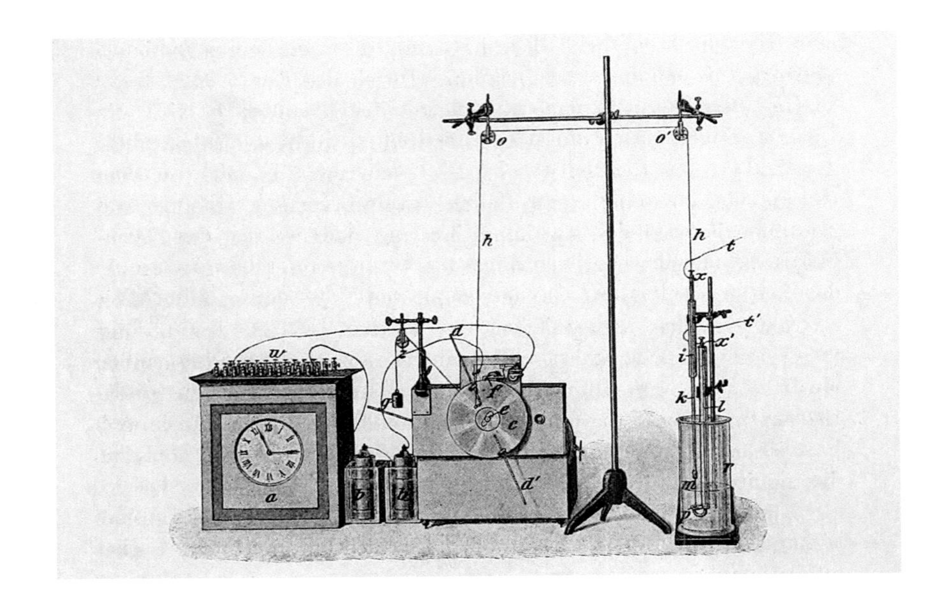

transferred to the imaging process in the time-lapse. Or, vice versa, as if the mechanisms able to register the motion inside the frames of the image could shape and constrain the movement of the plant in a controlled environment. In this case, the material culture underlying Pfeffer's images unveils a technical background of tools and instruments geared to the controlled modification of environments that surface both in the images and the altered plants. The material epistemology of these assemblages of light and plants is a fascinating element for us, one that will be developed in later chapters.

2.9

Wilhelm Pfeffer's equipment for an automatically adjustable illumination system using a clinostat among other devices; from Pfeffer, *Untersuchungen über die Entstehung der Schlafbewegungen der Blattorgane* (Leipzig: B. G. Teubner, 1907), 295.

THE LICHTGENUSS OF THE IMAGE AND THE EXHAUSTED LANDSCAPE

Drawing on analogies between images and vegetal sensitivity, Russian plant physiologist Timiriazev, a contemporary of Wiesner and Pfeffer, acknowledged that "the life of a plant is like a phantasmagoria, a successive series of changing magic-lantern pictures."[87] We have shown how Wiesner's work went beyond mere analogy and started using photographic techniques to characterize the behavior of plants quantitatively. Techniques such as his surface photometry were not the only instruments to register the behavior of plants. Moreover, they were instruments that helped to define plant perception. They provided a set of practices parallel to such contemporaries as Harold Wager, who, working on plant perception, defined epidermal cells as "very efficient lenses."[88] As a result, in these setups, the formation of the photographic image and the vegetal organic surfaces concurred. The parallelisms we have shown between the time-lapse technique and the clinostat in Pfeffer's experiments exemplify how early moving images belong to material circuits linked to the control of environments that literally modify the shape of plants. These form part of the recursive chains of cultural techniques of living surfaces.

From the modern global logistics of vegetal matter addressed in the previous chapter, we have traveled to two different technonatural assemblages[89] in the German-speaking plant physiology research of the turn of the twentieth century. There, a cultural and technological articulation of plants and images took place and condensed into a cultural technique whose impact on the environment can be traced in agricultural programs, earth observation systems, and models of the earth still in use decades after the two world wars. In the exhausted landscapes of the twentieth and twenty-first centuries, however, the vitalistic—even *naturphilosophisch*—appetite for light is still pervasive: as imaging drones, planes, and satellites that shape the growth and movements of the environmental surface complex. Not that this link is direct, but our book traces these connections in deciphering key elements for understanding the significance of living surfaces as part of large operational loops of imaging and shaping.

3

PHOTOCHEMICAL PLANETARITY

Under the microscope's lens, vegetal matter started to look different: the view of chlorophyll encased inside transparent lenticular membranes helped also to explain photosynthesis. The plant's surface was understood to be covered with chloroplasts, fueled by the radiant energy of light and several chemical processes inside the plant. During the last decades of the nineteenth century, photometry merged vegetal and photographic surfaces as part of constructing a vegetal growth model that related to light intensity. The photometric sensitivity of the plant guided the formation of the new corpuscles: only certain thresholds in the intensity of light surrounding the plant triggered the formation of new corpuscles of chlorophyll. Julius Wiesner's photometric observations suggested that new vegetal cells were produced only when the intensity of light fell between upper and lower specific measured values. Had Ansel Adams's Zone System been invented in Wiesner's time, plants would have been described as zoning the landscapes of light, foreshadowing the photographic zoning of arable landscapes that we examine in later chapters.

The previous chapter's engagement with Wiesner's double view of the vegetal and the photographic surfaces was one stop along the way of the scalar adventure of this book, and now we turn to the surface of planet Earth. While the book is situated in the broad context of visual culture studies and media theoretical discussions, we are also informed by sources that relate to the shift in the biological and geological sciences to datasets, images, and calculated surfaces. While the story is far from linear—as if we could shift from visual depictions to data, from operations of light to operations of calculation so straightforwardly—it helps to structure our argument according to historical case studies that also bend the scope of mediations in ways that can intersect both operations of knowledge and operations of terraforming.

In this chapter, we turn to the planetary surface, as it was formulated in twentieth-century scientific work in the Russian Empire and, subsequently, the Soviet Union. More specifically, this concerns the famous work by Vladimir Vernadsky, who introduced the notions of the *biosphere* (admittedly, coined some decades earlier by Eduard Suess), the *noosphere*, and, subsequently, the *technosphere*.[1] We address Vernadsky's work not merely for purposes of historical investigation, as interesting as it would be, but as far as it concerns the broader scalar logic of images as surfaces and growing surfaces as nested in a history of technical mediations.

The decades between the late 1800s and the early 1900s were characterized by rapid colonial expansion, massive resource grabs, and the establishment of a global food system that Hannah Holleman, in her work on the Dust Bowl, has described as a "new imperialism."[2] While standard accounts of this period have emphasized the importance of technological developments in communications, management, and transport as driving forces of the acceleration, scholarly work from different perspectives has emphasized the crucial role played by the scale of the territories involved in developing these technologies. For example, in the 1940s, Siegfried Giedion described the experience of seeing the cultivated landscapes of the American Midwest from a train. He writes, "This elimination of time, together with the mystery of dimension, produced the mechanization of agriculture."[3] More recently, Deborah Coen has shown how imperial access to vastly different territories enabled the birth of climate sciences and mapping scales and differences across regions in the Habsburg Empire.[4] A similar realization applies to the Russian Empire: the vast landscape and access to very different regions enabled the emergence of different comparative perspectives and research possibilities, including some that arose for very practical reasons of maintaining the empire. The focus on the study of soil by Vasily Dokuchaev is a good example of the interest in mapping the extent of the steppes and the variations in their geography.

His work gained further institutional support when he studied important critical events such as the harvest failure of 1891. This, subsequently, led to the maturing of soil sciences, as Dokuchaev's expeditions evaluated "the soil and natural resources of Nizhnii Novgorod province, in the north of the black earth region, and Poltava province, in the heart of the black earth region in Ukraine."[5]

The steppes, including those in Ukrainian territory, had been turned into agricultural land at the same time as being captured as visual landscapes in nineteenth-century painting, which codified them as part of the Russian Empire. In some ways, we see how plowing and (imperial) visual production went hand in hand with expeditions in the 1880s and 1890s that observed the constitution of the fertile, moist black soil. Furthermore, from the perspective of the Russian war on Ukraine that started in 2022, we must be aware of the long imperial politics of the state, part of which is found in this scientific story of how the planetary living surface became defined—englobing, in its definition of fertile soil and farmlands, those in Ukraine—or controlled, as in energy networks such as gas pipelines. Finally, the broader context of the discussions in the late nineteenth century recalls many of the current practices and discourses around climate change and land use: deforestation is suspected as a cause of environmental issues, with afforestation as one often suggested remedy, alongside other projects for systems of artificial irrigation.[6]

Within this lineage of thought, we propose to read the narrative and scientific description of planetary envelopes and surfaces that feature in Vernadsky's writings and beyond. While there is a lineage from Dokuchaev's research on soil (the black earth), cultural techniques of plowing and grazing, and other terrestrial forms of "terraforming," Vernadsky's perspective emerges in the context of further industrialization. In other words, Vernadsky's work has a foot in earlier soil research, but it also emerges against the backdrop of the cultural technical resurfacing of the world through chemical industries. In his own words, the First World War "radically changed my geological conception of the world."[7] The recently recurring terminology of "planetary scale" revives some of Vernadsky's ideas. One is tempted to add that the continuity of chemical industries and the weaponization of biosphere dynamics are crucial contexts in the twenty-first century. We want to approach Vernadsky's work as highlighting the importance that the planetary scale had in the elaboration of a biogeochemical model of the living. In this task, we understand scale as a perspective for critically analyzing a process of mediation to unveil otherwise invisible aspects and relations: it emerges both as a subject of research and as a methodological cue. Max Liboiron recently examined harm and violence produced by the pollution of microplastics: "Knowledge systems such as political ecology, cultural geography, and environmental justice are just some of the ways to look at how systems of value and knowledge animate relations. Scale is another."[8]

Scale features in this chapter in multiple ways: the planetary becomes an instance of mapping large-scale systems of precybernetic feedback loops. Such processes interact in ways that scale up and down across the dynamic boundaries of the planet. Here, "planetary" does not refer solely to a particular space but to the system of biogeochemical transformations at play. In addition, the planetary takes up the role of an interface: an epistemic handle on things beyond our immediate grasp. This has a political angle, too, when we read it in relation to the question of scale, for example as it emerges in Anna Tsing's note on the plantation logic of scale.[9] This mode of scaling subsumes specificities into large-scale units, cultivates particular kinds of modular worlds, and becomes an epistemic operation with a particularly violent history, too. In Tsing's words, "Scalability is, indeed, a triumph of precision design, not just in computers but in business, development, the 'conquest' of nature, and, more generally, world making. It is a form of design that has a long history of dividing winners and losers. Yet it disguises such divisions by blocking our ability to notice the heterogeneity of the world; by its design, scalability allows us to see only uniform blocks, ready for further expansion."[10]

Consider the point about scalability in relation to the subsequent discussion of Vernadsky's surfaces: Are we dealing with preparing ready-made units of knowledge that program and stamp the world according to its own image? With what kind of scalar inscription is Vernadsky operating? While we focus primarily on his work in this chapter, we must note that these questions go far beyond Vernadsky. Hence, we also consider the question of scales in relation to parallel studies of the period, such as the thermodynamic-based work of US biophysicist Alfred Lotka, summarized in his book *Elements of Physical Biology* (1925). Lotka boldly addressed "the entire body of all these species of organisms" as a "World Engine," and characterized it as "a vast unit, one great empire."[11] These sorts of scalar operations visible in rhetoric and scientific method prompt us to ask: What empires do scaling operations establish? What forms of supposedly natural growth are already included in calculations and projections of empires and other units of measurement? Any discussion of scales bends in many ways: some have imperial aims and some aim to facilitate the understanding of the complexity of surfaces that incorporate more than spatial dimensions. To quote Liboiron again: "Scale is not about relative size" but about "what relationships matter within a particular context."[12]

Reading Vernadsky's work and concepts regarding dynamic surfaces presents a significant case of planetarity far beyond its own scientific context in biogeochemistry. The model of the planet as a cosmic entity where the thin critical zone of life enveloped in dynamics from bacterial populations to atmospheric chemistry becomes the synthesizing site of different scales of surfaces that we are interested in mapping. Bruno Latour went so far as to claim that it is this "biofilm, a varnish, a skin, a few infinitely folded layers," where all the stakes for the terrestrial dynamics of knowledge, as well as their bespoke politics, lie: "Speak of nature in general as much as you like, wonder at the immensity of the universe, dive down in thought to the boiling center of the planet, gasp in fear before those infinite spaces, this will not change the fact that everything that concerns you resides in the minuscule Critical Zone. This is the point of departure and return for all of the sciences that matter to us."[13]

As such, this cartography of multiscalar planetarity not only echoes the more recent Gaia and Anthropocene discussions but it also connects to contemporary issues of planetarity from media studies to speculative design, without forgetting the context of environmental sciences. The link to Anthropocene discussions is relevant in terms of how the earth systems sciences are one version of a longer history of understanding the interconnected dynamics of different planetary forces, where knowledge of and through surfaces (such as remote sensing and satellite imagery of geographical landscapes) plays an instrumental role.[14] As Lynn Margulis and Dorion Sagan put it in their popular science book *What is Life?*, "Vernadsky dismantled the rigid boundary between living organisms and a nonliving environment, depicting life globally before a single satellite had returned photographs of Earth from orbit."[15] The envelopes, films, and other spatial concepts for solar-driven energetic transformation of the living matter addressed by Vernadsky that we discuss in this chapter are examples of planetarity read from the perspective of its growing surfaces and "the slow penetration into the Earth of radiant energy from the sun."[16] In other words, even before the actual cosmic perspectives enabled by space technologies, Vernadsky's reading and insights about the earth were cosmic in how they included the key driving force of living matter that was not present in the theories of physics or mechanics.[17] This view was material, even empirically concrete (even if full of abstractions, too), while it referred to a surface that was not just geophysical but full of biogeochemical processes from soil to the atmosphere and that bundled up into a view of geopolitics that is even more prominent now, in the 2020s, with energy and land use being central components of distribution of power.

THE BIOSPHERE

Let us rewind to the 1920s and the most central term in this chapter. In 1926, Vernadsky published *The Biosphere*, one of the first attempts to describe the ensemble of living processes on the earth as a holistically dynamic. Scientific domains such as geology, physics, and chemistry had already probed into the problem of creating models of the entire planet, producing their views as a result.[18] *The Biosphere* was Vernadsky's proposition to analyze life at a similar scale but, in his words, not as an abstract geometrical model or as inert matter.[19] His project aimed to describe the domain of biochemistry—the transformation of matter and energy as they link to questions of life—through a set of laws analogous with those of physics, though not irreducible in the same way. To do this, observations and data all around the planet enabled the interactions between living and nonliving entities to be quantified and compared from the point of view of the circulation of energy and the movement of matter. *The Biosphere* was the result of this project, and biogeochemistry was the practice that was perceived to be able to account—at the same time—for considering biological, chemical, and geological phenomena at the scale of the planet.

The term biosphere had been coined earlier, though, when it was introduced in 1875 by the Austrian geologist Eduard Suess. In 1899, the oceanographer Sir John Murray pointed to the multiple coatings of planetary phenomena, including the notion of geosphere:

When we regard our globe with the mind's eye, it appears at the present time to be formed of concentric spheres, very like, and still very unlike, the successive coats of an onion. Within is situated the vast nucleus or centrosphere; surrounding this is what may be called the tektosphere, a shell of materials in a state bordering on fusion, upon which rests and creeps the lithosphere. Then follow hydrosphere and atmosphere, with the included biosphere. To the interaction of these six geospheres, through energy derived from internal and external sources, may be referred all the existing superficial phenomena of the planet.[20]

Gradually, geospheric thinking spread far beyond the initial sciences to contexts as varied as meteorology and geophysics, where terms such as the troposphere, stratosphere, and asthenosphere were coined to designate different layers inside and outside of the earth.[21] While Suess understood the biosphere as the planetary envelope that encompasses all life on earth, Vernadsky added to this definition an interfacial character between the earth and the cosmos: it became "the envelope of life where the planet meets the cosmic milieu."[22] In contrast to Suess's descriptive under, Vernadsky's definition entailed an organized approach to the activity of the whole as if it were a self-regulated entity in continuous exchange with its surrounding environment.[23] The biosphere became understood as an active geological layer, which thus involved a central methodological problem in relation to the notion of scale and its epistemic implications: when dealing with the planet as a whole, individual phenomena had to be reframed at a larger scale to account for the cascading series of interlocking dynamic patterns that could not be understood merely locally. In his words:

Historically, geology has been viewed as a collection of events derived from insignificant causes, a string of accidents. This of course ignores the scientific idea that geological events are planetary phenomena, and that the laws governing these events are not peculiar to the Earth alone. As traditionally practiced, geology loses sight of the idea that the Earth's structure is a harmonious integration of parts that must be studied as an indivisible mechanism.[24]

In the context of the earth's structure, understood as a "harmonious integration of parts," *The Biosphere* defined the "planetary" as the scale at which the relationships between biochemical phenomena and geology mattered. The study of the planet required it to be addressed as a "holistic mechanism."[25] Despite the explicit reference to a "mechanism," his work should not be linked to a mechanistic representation of life, which he explicitly rejected.[26] It is instead an all-encompassing metaphor that describes a system of linked phenomena which, in opposition to a multiplicity of "essentially blind"[27] accidents, sense each other and make sense of the presence of a different scale, where scale "permits us to make sense of mediated, extended, and projected experience."[28] Geological events, including such as those linked to biological life, were suggested as connected to each other, giving rise to a dynamic system in equilibrium. In other words, *The Biosphere* needed to be understood from the central position of the scale of the planet in relation to the mediating role of the practices of *biogeochemistry*.

The biosphere envelops the planet as its uppermost layer, from a depth of a few kilometers below the sea level up to the troposphere's limits. In current environmental sciences, this is also referred to as the thin sphere of the Critical Zone itself, often defined through its heterogeneous multiscalarity, "ranging in scale from the mineral-water interface to the globe."[29] No forms of life are known outside this thin sphere, and barely any knowledge of the planet could be experimentally produced outside this zone. As chemist Frank W. Clarke put it in his influential *The Data of Geochemistry*, "Our knowledge of terrestrial matter extends but a short distance below the surface of the earth, and beyond that we can only indulge in speculation. The atmosphere, the ocean, and a thin shell of solids are, speaking broadly, all that we can examine."[30] Later, ocean-floor research, as well as space travel, might have changed parts of this claim, but the basic point is still valid when it comes to the biospheric context of knowledge as a keyhole through which broader planetary and interplanetary systems are understood.

While this merger of the living and the knowable would give rise to Vernadsky's later work on the noosphere, Clarke's reference to "all that we can examine" was already included in the scope of *The Biosphere*. Vernadsky's new science of biogeochemistry was also to move beyond any single discipline in its holistic take on such dynamic phenomena. It was to be grounded in results from different fields; Vernadsky proposed to rely on a series of "empirical generalisations," which allowed him to introduce principles and distinctions needed to elaborate his model and to strengthen its scientific validity in the eyes of his peers.[31] In order to be able to develop his arguments, he required the formulation of such statements, arguing that they were not introduced as theoretical hypotheses. They were formulations grounded in induction, well established by experience, which could not be fully proved due to the scale and variety of their domain. Throughout *The Biosphere*, he postulated a series of such positions and claims, among them, for instance: the permanent difference between inert and living matter, the presence of life during all geological periods, and the invariability of the chemical influence of living matter. In the context of this book, one of these—the ability of plants together with autotrophic bacteria to transform the energy of the sun and provide the other beings and processes on the earth with free chemical energy—is particularly relevant to our argument. This is what he named "the cosmic function of plants," an empirical generalization that worked as an interscalar vehicle where the cosmic scale was brought into the molecular level of photosynthesis.[32]

THE COSMIC FUNCTION

Immersed in space, the planet receives incoming electromagnetic flows. In the uppermost layer, they are transformed into free terrestrial energy. As an early version of the trope of Spaceship Earth, Vernadsky's biosphere, absent from other known planets, is an active envelope, which in the encounter with the surrounding "cosmic force" differentiates the flows into a variety of chemical, mechanical, and molecular forms of work:

> A new character is imparted to the planet by this powerful cosmic force. The radiations that pour upon the Earth cause the biosphere to take on properties unknown to lifeless planetary surfaces, and thus transform the face of the Earth. Activated by radiation, the matter of the biosphere collects and redistributes solar energy and converts it ultimately into free energy capable of doing work on Earth.[33]

Concerning such phenomena of life at the scale of the entire planet, matter is "activated" by the cosmic radiation that reaches the earth, which is mostly light coming from the sun. This process of activation by light recalls the photochemical domain of phenomena that we examined in the previous chapter. In this broad context, practitioners such as William Henry Fox Talbot wrote about light exerting an action that caused changes in material bodies while investigating whether these changes could be made visible on surfaces such as paper.[34] Later, in Vernadsky's work, vegetal surfaces made visible the action of light on the surfaces of the earth: "In the impact of a forest on the steppe, or in a mass of lichens moving up from the tundra to stifle a forest, we see the actual movement of solar energy being transformed into the chemical energy of our planet."[35] The growth of forests or the spread of lichen acted as evidence of one form of the activation of the matter referred to by Vernadsky. However, one might ask: What is the relation between photosynthesis and such an abstract notion of activation of matter in the biosphere? Also, taking into account the discussion from chapter 2, is Vernadsky's idea of activated matter related to a model of calculated photosensitive surfaces akin to the photometric surfaces used in photography?

To answer the first of these questions, it is important to acknowledge that Vernadsky described plants, together with autotrophic bacteria, as the only living beings able to transform incoming cosmic radiation into chemical energy. In his words, "All living matter can be regarded as a single entity in the mechanism of the biosphere, but only one part of life, green vegetation, the carrier of chlorophyll, makes direct use of solar radiation."[36] The biogeochemist, emphasizing the importance of plants in the order of the biosphere, established a parallelism between their role and the "cosmic force" of light by addressing the plants' photosynthetic singularity as their "cosmic function."[37] This phrasing had been used before: Vernadsky alluded with it to the title of a well-known lecture given in 1903 by Russian plant physiologist Kliment Timiriazev, "The Cosmical Function of the Green Plant." In the lecture, Timiriazev referred to the oxygenation of the atmosphere as the cosmical function of the plant. That is, he named the terraforming character of plants that we discussed in chapter 1 in relation to the cultural technique of experiments with plants in transparent cases in enclosed environments. In the earlier discussion, the recursive character of the cultural technique mediated between the different globes concerned, from the glass jar to the Wardian case to the planetary globe. In this chapter, the cosmic force arrives from outer space to become the energy-transforming ability of a plant. Such a biogeochemical model of life proposes that the cosmic perspective involves a reflexive recursion; in Katherine N. Hayles's words: "Reflexivity is the movement whereby that which has been used to generate a system is made, through a changed perspective, to become part of the system it generates."[38] This recursion among scales abstracts plants as vegetal surfaces, vegetal surfaces as living matter, and living matter as an activated geologic coating responding to the cosmic energy of the sun.

SCALING MATTER

The Biosphere was written between the First and Second World Wars. As Etienne Benson has observed, the supply chains and the international networks of finance, trade, and communications of the second half of the nineteenth century were ripped apart during the First World War. This led to a widespread interest in the question of self-sufficiency at the scale of the nation-state—that is, the problem of territorial autonomy in relation to food, energy, and manufacturing systems—expanding a long-standing agricultural question. It was important to acquire reliable knowledge of available resources and to develop a logistical understanding of "their accessibility and value in relation to changing levels of supply and demand, the availability of substitutes, new methods of extraction and processing, and economic and geopolitical constraints."[39] In Benson's words, "Ecologists, demographers, and geochemists applied techniques and concepts developed to manage strategic materials during the war to the study of exchanges of matter and energy between organisms and their environments."[40] Before writing *The Biosphere*, Vernadsky founded an institute devoted to the inventory of Russia's strategic materials, even proposing the completion of an "international radiography of the earth's crust."[41] He also advocated for a nationwide program devoted to research on radioactivity.[42] The metabolic circulation of matter and energy on a global scale remained a central part of Vernadsky's work, continuing earlier developments of the bioenergetic notion of metabolism, such as were proposed by the physician Julius R. von Mayer and chemist Justus von Liebig in Germany, and the Ukrainian physician Sergei A. Podolinsky.[43]

We want to pay special attention to the scale of the planet that emerges in bioenergetic notions of metabolic flows. *The Biosphere* can be described in terms of Zachary Horton's theorization of scale as both a producer of difference, which gives rise to stable entities, and a milieu of relational dynamics between them.[44] In order to elaborate the model of the interaction between the cosmic radiation and the surfaces of matter on the planet, Vernadsky established as an empirical generalization a central distinction: "Two distinct types of matter, inert and living, though separated by the impassable gulf of their geological history, exert a reciprocal action upon one another."[45] Living matter encompassed the total sum of living organisms, while inert matter referred to the ensemble of solid, liquid, and gaseous formations that envelop the living, such as soils, oceans, rivers, and the atmosphere. The idea of "activated matter" operated in two senses: first, as we have just seen in relation to the cosmical function of plants, as a diagram of living matter, where radiation is transformed into free chemical energy; second, in a logistical manner, as a "conveyor and a storage of cosmic energy,"[46] circulating it and retexturing with it the inert shells of the biosphere. This version of a planetary conveyor belt consisted of living and inert matter, as well as the logistical processes and circuits among them where energy and matter were transferred in both directions.[47]

Furthermore, the planet was modeled as a surface. Vernadsky's research program unfolded through a vocabulary that flattened the differences inside living matter to consider it instead as a continuous surface: "Living matter clothes the whole terrestrial globe with a continuous envelope."[48] Emphasizing this surface condition, Vernadsky addressed assemblages of vital matter at a smaller scale as "living films." Such films were, for instance, the continuous layers of green life (green autotrophic organisms) that cover the surface of oceans alongside phytoplankton or, in the case of land, "one living film, consisting of the soil and its population of fauna and flora."[49] While mostly continuous, this film was however remarkably thin. Vernadsky added:

> Only some tens of meters above the surface in forest areas; in steppes and fields it does not reach more than a few meters. . . . The living film thus covers the continents with a layer that extends from several tens of meters above ground to several meters below (areas of grass). Civilized humanity has introduced changes into the structure of the film on land which have no parallel in the hydrosphere. These changes are a new phenomenon in geological history, and have chemical effects yet to be determined. One of the principal changes is the systematic destruction during human history of forests, the most powerful parts of the film.[50]

What is interesting is that this film is not merely an empirical description, it has become an aesthetic and epistemic device for understanding scales. It includes a way to characterize the distribution of different vegetation (and other living) formations, while it becomes an abstract notion holding the planetary reference in place. To add, the tension between the differences inside the ensemble of entities encompassed in such films and the abstract and unifying character of the notion of the surface itself defines, in Horton's terms, scalar media: "A 'scale' is a singular resolution of ontological difference between two surfaces."[51] Here, Horton's focus includes the operational rendering of complex formations into surfaces that can also encode depths or, in the case of vegetal surfaces, all sorts of photosynthetic and energetic processes. Indeed, while paying attention to examples such as films, Horton's definition of media as a "series of machinic differential operations within [a] cosmic-historical flow of matter-energy"[52] further underlines the connection to fundamentally material kinds of surfaces that are instrumental in recursive operations of scaling. Indeed, the scalar media of living films, patches, and vegetal zones become way of underlining the stakes of planetary mediations. Furthermore, these mediations depend on not just what scale is identified as relevant but how this scale embodies, counts, resolves, dissolves, negotiates, and amalgamates relationships.[53]

THE PHOTOMETRIC PLANT AS SCALAR MEDIA

Once the living is abstracted as flat surfaces of living matter, *The Biosphere* turns to the operations of energy of sunlight. We have already seen how important plants are for Vernadsky, who described them as having a "cosmic function." Earlier research in plant physiology had a significant impact on his thinking, not least the work of Timiriazev and Wilhelm Pfeffer.[54] However, the influence of Julius Wiesner has remained underestimated. In other words, the research by Wiesner that we discussed in the previous chapter can also be identified in Vernadsky's ideas. By engaging with this connection, we will see how the cosmical function of the plant appears not only as a recursion within the multiscalar model but also as a significant case of scalar media included in Vernadsky's arguments. The photometric growth as a model for the surface of the plant in Wiesner's work becomes, for Vernadsky, a model for the logic behind the distribution of the surface of living matter that envelopes the planet. Whereas Wiesner described the structure and growth of the "living substance" in the plant in detail,[55] Vernadsky addressed instead the scale of biogeochemical multiplication of "living matter" on the earth. The "movement of life"[56] in relation to the presence of light was mapped by Vernadsky following Wiesner's model of the growth of the plant. As a consequence of this epistemic mapping, it was also rendered calculable. The photometric surface of the plant becomes a form of scalar media of the biosphere. Plant life is scaled up to the planetary in order to arrive at a biogeochemical model of life on earth.

In order to see this point more clearly, we should recall how, right after his initial description of living matter as a biogeochemical assemblage of entities and processes, Vernadsky turned to the unique role of plants as and on living surfaces. He subsequently explained how several studies, including those by Wiesner, had shown these organisms had evolved and adapted to the cosmic function. In particular, Wiesner's idea that light "exerted a powerful action on the form of green plants" was emphasized in *The Biosphere*.[57] In order to underline that the form of the vegetal is affected by light, Vernadsky quoted the Austrian plant specialist: "One could say that light moulded their shapes as though they were a plastic material."[58] He postulated this behavior again as an empirical generalization: "The green apparatus which traps and transforms radiation is spread over the globe, as continuously as the current of solar light that falls upon it."[59] Moreover, if plants inside transparent cases had been shown to actively produce the necessary gaseous equilibrium for their survival, the surface of living matter on top of the earth was also able to autonomously regulate the gaseous mixture in the atmosphere. The photochemical morphogenetic dimension of the interweaving between vegetal matter and the rays coming from the sun provided him with an explanation for the efficient spreading of life all over the terrestrial surface.[60]

The influence of plant physiologists can also be witnessed when Vernadsky wonders about the cause of plant-light interaction. Do plants grow and adapt their shape to a passive background of light or, conversely, does light sculpt them as if it has invisible fingers? This was exactly the old and polemical dispute between Darwin and the German physiologists in relation to the sensitivity of plants that grounded the work on photographic plant growth that we discussed in the previous chapter.[61] Following what Wiesner had proposed two decades earlier in his research on phototropism, Vernadsky suggested a synthesis of sorts: "The solution should probably be sought in a combination of both approaches."[62] Wiesner, as we have shown previously, explained how the shape of plants must be understood as an adaptation to their active environment of light. Neither accumulations of vegetal matter reacting to external forces nor free organisms with the ability to experiment with their own form, plants developed leaves where the intensity of light equaled their Lichtgenuss. This feature of plants was scaled up by Vernadsky to the planetary: "The firm connection between solar radiation and the world of verdant creatures is demonstrated by the empirical observation that conditions ensure that this radiation will always encounter a green plant to transform the energy it carries."[63] Like individual plants, whose leaves are placed where the intensity of light coincides with the value of their Lichtgenuss, the living surface is presented in a dynamical equilibrium with light in such a way that every incoming ray ends up meeting the appropriate creature that efficiently metabolizes it.

As discussed in chapter 2, in the context of this experimental plant physiology, vegetal growth was measured and modeled through cameraless photographic techniques. In the already mentioned conference on the cosmical function of plants, Timiriazev discussed their photographic character: "Chlorophyll plays in the living organism the part of an optical sensitiser."[64] In a similar vein, Wiesner provided a photochemical model for plants' ability to deploy their surfaces against the varying environment of luminous conditions. Likewise, Vernadsky proposed a biochemical model to explain the capacity of living matter to spread and perform its photosynthetic function everywhere on the earth. The green apparatus, the latent disposition to transform the rays of light into the chemical energy necessary both to sustain and expand the living, was spread as an active interface all over the surfaces of the earth. In other words, Vernadsky took note of the work of a previous generation of plant physiologists on the interaction of plants with light and scaled it up to the planetary. He postulated that the ensemble of all life forms could be understood as a holistic entity tied together through the metabolism powered by the energy of light. He named this totality *the living matter* and characterized it as a surface in continuous formation, as if it were a vegetal organ adapting its shape in relation to the light. The photographic character of the surface of plants that resulted from turn-of-the-century continental plant physiology was brought to the planetary by Vernadsky. By doing this, life on earth was modeled as a photographic development of the light of the sun or, in other words, as an image of chemical energy in continuous formation—as if developing an ever-changing photosensitive film.

While living matter grounded the oxygenation of the planet, they simultaneously refashioned its uppermost geologic crust: "Seen from space, the land of the Earth should appear green."[65] The carpentry of living on the planet became an interfacial layer that shifted from the metaphorical face of the earth (such as in Suess's vocabulary) to a technique of registering and processing cosmic rays. All these different examples made the media-specific and technical issue of the sensitivity of the photographic surfaces into a generalized feature of living surfaces in an energetic ontology of light.

TECHNOLOGIES OF LIFE

To understand the significance of the change of scale in the approach to life and the nuances of taking the surface of the plant as well as vegetal formations as scalar media, it is worth observing that the notion of living matter was outside the scope of the research carried out by most of

Vernadsky's contemporaries. At the beginning of the twentieth century, living matter was conceptualized as an abstraction, where life was understood outside the standard categories of species and individuals operating in the disciplines linked to biology.[66] Aware of this, in a conference two years after the publication of *The Biosphere*, Vernadsky emphasized that his work had proposed a completely different (biogeochemical) model of life. In his words, understanding life as a planetary phenomenon involved "fundamental conceptions of biology," mostly focused on morphological characteristics of species, which had to be "submitted to radical modifications."[67] The scale of the entire planet was brought to the foreground and, as a result, the distinction between the living and the inert at that scale could be understood only by using a series of new knowledge tools, different from those taken from biology. This already, in some ways, hinted at the later theorization of the noosphere as a *bio-techno-sphere*, where "man and its exosomatic instruments, the earthly environment, and all technologies became inseparable."[68]

By the late 1920s, Vernadsky's materialism had grown to be similar to the philosophical critiques of hylomorphism featured in the work of Alfred North Whitehead and Henri Bergson, which acknowledged the constitutive links of organisms with their environments. Their names feature explicitly in Vernadsky's writing. As Vernadsky emphasized, organisms cannot be understood and should not be studied as independent from their living environments, nor should the two be opposed either: organisms and environments are tightly coupled in a dynamic ontology of living and nonliving matter.[69] For Vernadsky, though, the focus was on proposing new ways of understanding biogeochemical phenomena as *technologies of life* that were operating at the back of the *biogenic migration* of living matter.[70] While his position was incurably teleological in how he read paleontological observations as proof of a unified, determined direction of evolution, it still afforded interesting, even radical, ideas about originary technicity that resonated as part of the broader media cultural enthusiasm not only with new technologies of early twentieth century but with their "natural" counterparts too.[71]

Understood under the umbrella of such a technology of life, for example, the biogeochemical view of a "swarm of locusts" depicts them not only as a biological species but as exoskeletons made of minerals:

geological matter animated, "extremely active chemically, and found in motion."[72] Living and inert matter were found to be always in circulation, where the observation of the movement of the former anticipates the distribution of the latter and vice versa.[73] Such a position provided a radical ontology of movement that allowed such seemingly mistaken interdisciplinary jumps to emerge that also saw geology as part of living surfaces. The Russian geologist Andrei Lapo, one of the early specialists of Vernadsky's work, points to how fundamental this position was to the model of planetarity that Vernadsky presented: "Living matter is a specific kind of rock . . . an ancient and, at the same time, an eternally young rock. A rock which creates itself and destroys itself to originate again in new generations in the innumerable forms constituting it."[74] In other words, this theory of the planet as a field of capture of cosmic energies connected the earth-sun system to the reshaping of the earth's crust and strata, with mineral and gaseous consequences beyond living matter, including fuels, carbonate, and phosphate deposits, soils, atmospheric gases, and so on. It is as if the radiation of the cosmos, thanks to these technologies of life, left its print on the earth, a dynamic geological print made up after the biogeochemical activation of the surface by the living films on top of the planet. One could refer to these large-scale prints, traces, and marks also as *autographic visualizations* that remain as one form of an archive of environmental changes.[75]

While we pay attention to such models of light and inscription of surfaces, at the back of Vernadsky's argument was also an "arithmetic point of view."[76] This approach to the biosphere required a way to deal with the living and the nonliving at the same time, without assuming either a mechanistic reduction or a vitalistic position, both of which he disregarded as "alien to science."[77] As an alternative, Vernadsky abstracted the model of gaseous diffusion and turned it into a general archetype of biogeochemical movement. "The diffusion of life is a sign of internal energy . . . and is analogous to the diffusion of a gas."[78] Living matter spreads like a gas thanks to its ability to grow by multiplication. In fact, multiplication was precisely the key feature for Vernadsky that distinguished living from inert matter. He described both processes of diffusion, gas, and living matter in the same terms as fluidly overcoming obstacles and producing pressure in the surrounding environment:

The dimensions of the planet also impose limitations. The surfaces of small ponds are often covered by floating, green vegetation, commonly duckweed (various species of Lemna) in our latitudes. Duckweed may cover the surface in such a closely packed fashion that the leaves of the small plants touch each other. Multiplication is hindered by lack of space, and can resume only when empty places are made on the water surface by external disturbances. The maximum number of duckweed plants on the water surface is obviously determined by their size, and once this maximum is reached, multiplication stops. A dynamic equilibrium, not unlike the evaporation of water from its surface, is established. The tension of water vapor and the pressure of life are analogous.[79]

In this universe of life and numbers, he also described a "speed of transmission of vital energy,"[80] which was meant to estimate, numerically, the intensity of the reproduction of the living surface in a specific place.[81] Mimicking what physicists had done with the kinetic theory of gases—the statistical approach to gases and their thermodynamics—he even defined the "internal energy" of this spreading and calculated it as the sum of "the separate energetic movements of its component particles."[82] The spreading of the surface of the animate layer was modeled as if it had acquired the characteristics of a gaseous substance. Vernadsky transferred the statistically measured architecture of the gaseous envelope to the behavior of life at the scale of the biosphere. That is, the living was not reduced to the physical so much as it was transported by it. Broadcast as a gaseous signal, it enveloped the globe and exerted pressure on its obstacles. And beyond that, it could be appraised by calculation. Even in the material world, numbers provided the backbone for how growth, reproduction, distribution, and other forces could be described without losing any sense of their vitality as active dynamics of planetary scale "terraforming."

A CHEMICAL RECOATING

The world was anyway radically geoengineered around the early twentieth century. Chemical industries recoated so-called natural surfaces, adding a further twist to the notions of geospheres that had been circulating for some decades. As Esther Leslie shows in *Synthetic Worlds*, the chemical industry founded an empire of analogs and replacements.[83] Manufactured materials as varied as aniline-based colors, plastics, celluloid, surface coatings, and synthetic oils took the place of organic originals such as natural pigments, ivory, or bones. Vernadsky was aware of this change, as he reflected in the mid-1940s: "Chemically, the face of our planet, the biosphere, is being sharply changed by man, consciously, and even more so, unconsciously. The aerial envelope of the land as well as all its natural waters are changed both physically and chemically by man."[84] His work on the "movement of life" inevitably connects *The Biosphere* to the parallel chemical gasification of agriculture brought about by corporations. For the sake of periodization, let us note that *The Biosphere* was published the same year IG Farben was founded, the infamous cartel that gathered the biggest German chemical corporations of the moment, including BASF, AGFA, and Bayer.

Crop dusters and other fogging techniques were not yet present in everyday agriculture in the early twentieth century, except for some early experiments during the 1920s in the United States. The Haber-Bosch process that synthesized ammonia as a fertilizer had, however, already been developed and introduced on an industrial scale in 1913. Since the mid-nineteenth century, following an agricultural crisis due to the depletion of soils by intensive practices, there was full awareness that the fixation of nitrogen was essential for the growth of plants.[85] Nitrogen could be extracted from the vast deposits of guano in South America and added as manure to fields. This improved the yields considerably, and it also fueled the development of "the new agricultural chemistry" initiated by the German chemist Justus von Liebig.[86] As nitrogen is the most abundant gas in the atmosphere, many attempts to extract it directly from the open air followed. These were, however, unsuccessful. The situation changed in the first decade of the twentieth century with the development of a process by Fritz Haber and Carl Bosch. Nitrogen-based fertilizers could be produced in factories. As a result, the use of synthetically fixed nitrogen spread immediately after its commercialization. Furthermore, it could be used in the production of explosives as well as in chemical warfare.[87]

Since the early twentieth century, the air has been artificially circuited to the soils of the planet in the form of synthetic nitrogen. The

Haber-Bosch process becomes then part of the recursive operations that link synthetic chemistry with the redefinition of growth and agricultural lands and, in the process, trigger large-scale repercussions for a variety of issues from geopolitics to climate change. As a matter of fact, after the Second World War, the growth of the use of these fertilizers was exponential; they even forced researchers to look for high-yielding varieties of crops, since the standard ones could not absorb the extra nutrients. Water consumption increased, and large-scale irrigation infrastructures were needed, together with different types of pesticides and their related techniques of fogging. Over several decades, this was marketed as the "Green Revolution,"[88] which refers to the ensemble of technoscientific developments, patents, management, and communication strategies that led to the planetary-scaled spread of industrial agriculture. The scale is huge: as chemist Paul J. Crutzen put it in his oft-cited paper on the Anthropocene, "more nitrogen fertilizer is applied in agriculture than is fixed naturally in all terrestrial ecosystems."[89]

To read Vernadsky's living matter in relation to industrial agriculture highlights a material and epistemic context that goes beyond plant physiology and other life sciences. It brings up a chemical background of operations that, interestingly, did not solely exist in the agricultural domain but also permeated the production of images. The planetary, as understood in this chapter and the book, started to include this mode of transformation of soil and surfaces as it was also connected to an entangled sphere of images. New chemical production was driving both kinds of surfaces. Photography, in particular, experienced a revolution thanks to the development of new and faster sensitizers.[90] With their aniline dyes, corporations such as Bayer, AGFA, and BASF transferred their industrial mastery of chemical cycles to shortened photographic exposure times and, by doing so, in turn expanded the operational space of photography itself. Measurements and scientific practices rely on chronophotography, for example the experiments discussed in chapter 2. Another clear case involves aerial photography: during the First World War, the new dyes gave rise to specific sensitizers that, as Michelle Henning says, "reshaped photography in response to the demands of aerial reconnaissance."[91] Sensitizers were developed to allow aerial cameras to see through the atmospheric haze. In this way, chemistry sensitized the exhausted soils of agriculture to produce faster developments of photosynthetic matter and increased the rate and range of images produced by photographic surfaces. Chemistry refashioned all the world's surfaces anew: photographic plates, soils, and, as demonstrated, vegetable surfaces, including agricultural ones. All kinds of surfaces emerged as central elements of knowledge and operations: first, planetary-scale life as envelopes of interactions of energy, and second, an experimental set of practices that fed into reforming those surfaces while also fixing new kinds of images to make sense of those surfaces. The recursive features of this operational sphere are both the key focus of our book and the site of epistemic inquiry that pertains to the environmental modes of interaction that have become again central in the past decades of debates, not least about the Anthropocene.

In Vernadsky's biosphere, as in Priestley's jar, a shell of inert matter encloses the living surface: gases above, rocks below. The photosynthesis of the green component of living matter requires a gaseous background to absorb and secrete its chemical sources and wastes. In Vernadsky's words, the movement of life can occur "only through a gaseous exchange between the moving matter and the medium in which it moves."[92] One could think of inert matter as a sort of infrastructure of the living, echoing Paul N. Edwards's words about nature as "the ultimate infrastructure."[93] The inert, however, needs to be understood as being adapted, transformed, and shaped by the living surface it hosts, problematizing any final difference between the two. (Furthermore, in the light of more recent developments in material sciences, Anthropocene research, and humanities—for example, in new materialism—it is impossible to claim that matter is just bluntly *inert*). This was, after all, one of the key points developed in this take on the planetary scale of biogeochemistry. In Vernadsky's words, "The organism deals with the medium to which it is not only adapted but which is adapted to it."[94] Like the plant that oxygenates the glass jar in Priestley's experiment discussed in chapter 1, living matter is, in Vernadsky's model, a "medium-forming" force[95] and requires an environment while it contributes to shaping and maintaining itself.[96] Matter is fundamentally reflexive in this material sense.

Even before *The Biosphere*, Vernadsky had perceived the soil surface as a fundamental scalar media of biochemical forces. Soil might be located physically as a thin layer on earth, but it was a biospheric force greater than its location, as it "wholly matches the huge active energy that is accumulated in soil's living matter and that is capable of transfer by soil penetrating gases."[97] This "medium" of negotiating scales concerned not merely the different spatiotemporal perspectives but a multitude of forces operating on living surfaces. As a model of planetarity, then, it presented a way to appreciate the patchwork of surfaces as reflexive, recursive operations that include multiple scales encoded onto a surface. Furthermore, it was also one variation on the discourse of "the green mantle" that Veronica della Dora skillfully tracks as an art historical and geographical theme. While the nineteenth- and early twentieth-century notions of green mantle found in Thoreau, Muir, and some botanical works "all express an understanding of nature as 'wilderness,' for example, as special areas to be protected by enclosing them within the sacred precincts of natural reserves and parks,"[98] here we have a multicolored and multiscalar patchwork in operation that is defined by biochemically created zones: a mantle, a texture, a cloth that connects across differences from changing ecozones to biomes, biotopes, and more. It represents a continuity in difference. The green mantle had its own version in earlier Russian soil science, too, some of which also included a strong nationalist element even while incorporating the global and the planetary as its reference points. Vasily Dokuchaev, the influential figure in Vernadsky's instruction whom we mentioned above, had referred to the soil as a *global natural object* that on the earth's surface was divided into "soil zones which blanket the entire globe, both the northern and southern hemispheres."[99] Dokuchaev described the earth as "multicolored ribbons of soil," where the color coding not only referred to different amounts of heat and light but also included a racialized reference to changes in human and animal (pigment) surfaces, "from white to grey, black, chestnut and copper-red."[100] With hints of Russian mysticism and nationalist undertones of (superior) soil, Dokuchaev's pedology functions as geopolitically tuned scalar media inscribed on the soil.[101] While soil mapping and coding became thus one form of articulating the reach of the empire and its reliance on logistics of food security that became a crucial issue after the 1891 famine, it was also an articulation of symbolic differences in terms of imaginaries of nation-states and empires: as a matter of fact, Dokuchaev observed the mistaken attempts to apply German agricultural methods to Russian soil while sketching out a proposal how to see these soil zones forming out of cosmic and planetary processes. According to Dokuchaev, the zones were defined by interactive processes between living and inert matter from water, air, and earth to plants and animals as *soil-formers*.[102]

Rather than a model of biological matter built on the resources of the inert, Vernadsky's *The Biosphere* can be read as the production of a synthetic layer of tools and conceptualizations that account for the mediating role of the planetary scale as it folds recursively onto surfaces and terrestrial systems. The planetary becomes not merely about the spatial scale of the planet but about the interlocking dynamics that fold the planetary in different materials, surfaces, and processes. Vernadsky's biosphere is about the transfer of scales and scalar logic, out of which the living and the inert dimensions of matter are imagined and brought back to the planet's surface. Usually, these themes of techniques of planetary scale are read through Vernadsky's idea of the noosphere. Still, as we have argued in this chapter, the environmental and political stakes were evident decades earlier than the actual emergence of the term.[103]

"Environments, like media," writes John D. Peters, "are delicate systems of contingent conditions for the organisms that live in them."[104] They constrain and modulate the forms of life they enable, and they are recursively transformed and even constructed in the process of that modulation. For example, the ozone layer is an interesting case in relation to the double bind between living and inert matter that characterizes Vernadsky's biosphere. It shows that inert matter can take a role as the infrastructure of the living, and at the same time, it shows how the living can be taken as the infrastructure of the dynamics of circulation and regulation of the (seemingly) inert shells of matter on the planet. Living and inert matter are interweaved so that, for Vernadsky, it is not even possible to state which came first. In other words, despite his insistence on differentiating inert from living matter, none of these two systems in *The Biosphere* is privileged,[105] substantiating what Bruno Latour has emphasized in relation to the Gaia model. There is no prominence, priority, or internal hierarchy among the systems in interaction.[106]

Throughout this chapter, we have argued that *The Biosphere* can be read in terms of surface tension between the organic and the nonorganic that is resolved in the gaseous model of living matter. Different dynamics of the circulation of matter—a logistical imaginary of planetary life—become the central driving engine in this model that also works with questions of light and energy. The model is linked to the parallel spread of the chemical industries, which also proposed a gaseous form of agriculture. New clouds started to appear at the back of the infamous industrial smoke clouds captured in Ruskin's poetic words in the nineteenth century.[107] Some decades later, new clouds were also related to new kinds of practices of growth, as well as extermination of life, both human and plant. What is more, this gaseous model of planetary transformation—chemical media or cultural techniques of chemistry—can be related to the visual technologies that had mediated, in the first place, the observation of plant growth. In particular, we have emphasized the infrastructural role of industrial chemical media in the photographic surfaces that allowed the advancement of the operational and measuring techniques of the cameraless experiments addressed in chapter 2. One needs the other to establish the other that comes back to support its foundations.

The techniques and infrastructures that constitute agriculture ultimately rely on the morphogenetic adaptability of vegetal life, its industrious persistence, and its preference for cycles. With minimum requirements, plants inevitably grow and spread. Any sort of modification, acceleration, or control is parasitic on this primary process of life.[108] Husbandry is not an activity geared toward making the plant grow—growth emerges within the plant itself—but is instead concerned with removing its obstacles. Doing this sets up a parasitic mode of relation: growth itself is enclosed, wrapped, enveloped, and measured. It is circuited in categories of scientific observation and classification that help to reoperationalize it. Now, it is not the plants anymore but the crops (including plantations) that grow. It is chemical additives that power their growth, and it is not only chemistry but a set of refined media techniques that model plant growth to close the loop between data-chemistry-crops-plants. This characterizes the different cases of hybridization between images and plants that we examine in the following chapters. But now, we have a sense of this scale of planetarity that comes to haunt not only modeling growth, light, energy, and their interaction but also the political issue of scale we referred to earlier with the help of Liboiron: what matters are the relationships bundled up in scalar media, the forces gathered in scalar operations including the focus on the planetary that is produced through specific cultural techniques of surfaces.

4

INNER COLONIZATION AND VISUAL AGRICULTURE

A woman sitting on the fence of a dacha occupies the center of the frame. She has her back to the camera, and our gaze is directed to a meadow extended onto a valley in front of her, in the distance. The country house, a reconstruction based on memories and photographs of the one in which the film director Andrei Tarkovsky spent long periods during his childhood, does not appear in the image. The scene at the beginning of Tarkovsky's film *The Mirror* (1975) emphasizes instead the act of looking at the meadow, which is observed twice: by the main character—the narrator's mother—and by the camera. The emphasis is not anecdotal; the meadow, like the house, is the result of a restoration based on memories, a recurring theme for Tarkovsky.

A field lay in front of the house; I remember buckwheat growing between the house and the road leading to the next village. It is very pretty when it is in blossom. The white flowers, which give the effect of a snow-covered field, have stayed in my memory as one of the distinctive and essential details of my childhood. But when we arrived to decide where we would shoot, there was no buckwheat in sight—for years the *kolkhoz* had been sowing the field with clover and oats. When we asked them to sow it for us with buckwheat, they made a great point of assuring us that buckwheat wouldn't grow there, because it was quite the wrong soil. Despite that, we rented the field and sowed it with buckwheat at our own risk. The people in the *kolkhoz* couldn't conceal their amazement when they saw it come up. And we took that success as a good omen.[1]

The description of the production of the scene presents the buckwheat as a surface whose growth occurs as part of a circulation of images. The image of a blooming meadow is first fixed in the memory of a child; years later, because of this memory, the land is rented, the existing oat and clover crops uprooted, buckwheat replanted, so that it blooms, again, visibly in the landscape. Finally, the image is fixed again, this time in the moving sequence on the film stock. In short, the image of the field circulates from memory to soil to film, depicting a transformation of the landscape along the way as it becomes planted and printed according to a memory.

From the point of view of Tarkovsky's work, this circulation of images and memories underlines the fact that the film was conceived as staging a process of remembrance. That is, the film deals with an evocation of spaces and moments of childhood, which instead of seeking an image of a lost time, focuses on the very elaboration of memory through images.[2] However, beyond Tarkovsky's specific interest in memory, it is important to note that the scene involved the interaction between images and an agricultural field. Moreover, it recalls what film studies scholar Graig Uhlin has argued about the Russian filmmaker's work, that it can be understood as "vegetal filmmaking."[3] Uhlin's argument starts from Tarkovsky's concept of the "vital process," also referred to as the "filmic inner state" or the "inner life of the shots," which refers to the internal rhythm of the sequences that allows them to connect with each other in an organic montage.[4] The filmmaker's view on montage can also be related to the radical surface condition of plants featured in chapter two: the environmental conditions that surround them merge into the shape of the plant.[5] Following Uhlin, this condition is what Tarkovsky's sequences aspire to: sensitive membranes through which invisible living processes surrounding the camera acquire a material form.[6]

More generally, the transformation of large areas to adjust them for the aesthetic needs of a film production are not unusual. For example, in order to emphasize selected textures, Michelangelo Antonioni painted green the famous lawn in *Blow-up* (1966), while the streets, doors, and even the fruits next to the industrial complex of *The Red Desert* (1964) were to be in grayscale. The material work on the surfaces, their visual texture, is an essential component in the fabrication of the filmic image as it becomes integrated into the architecture.[7] But this architecture is not limited to the urban conditions of built environments as projection surfaces or textures. In this sense, the introduction of agricultural techniques *as* cinema extends this *surface condition*[8] to the scale of the landscape: prairies and valleys prepared to become an image, crop patterns as more than backdrops, plantations as the imprints of industrial form of production. This is image- and filmmaking that tries not just to shoot images of such views but to merge as one, to become ecological in a way that captures a bundle of sequences, temporalities, and forces at play.[9]

This chapter argues that this interstitial surface condition of the meadow in Tarkovsky's scene relates to a wider domain of operations on landscapes that belong to contexts other than cinema. Such an aesthetic does not pertain only to the artistic expressions of films as they become large-scale prints of another sort, another economy. We look at what film studies theorist Nadia Bozak has called the cinematic footprint in the very concrete terms of impressions left on surfaces: the space of relationships, dependencies, and articulations between images, resources, and energy that has placed the twentieth-century production of images as one more of the industries under which the consumption and transformations of landscapes have taken place.[10] This footprint in Bozak's work involves not only the territories unearthed due to the fossil dependencies linked to the needs and the making of tools and film stock but "the subordination of nature as the root of industrial culture" entailed by the conflation of the categories of "resource" and "image."[11] The transformation of the plots as part of Tarkovsky's film production needs can be framed, beyond the author's intentions, within a broader framework of relationships forged throughout the twentieth century between images and landscapes. As we move beyond our opening example of Tarkovsky, this point becomes clear in an investigation of how agricultural lands and landscapes are fabricated both in colonial contexts and in terms of extractive industries. Here geographies of growth and irrigation become subsumed under this framing of image and land. As we will see next, these were processes that can be understood in terms of circulations of images—as in the buckwheat scene—insofar they took place in territories governed by an administrative gaze supported by visual means, such as aerial photography.[12] Such is the case of the major irrigation programs that transformed vast tracts of land in different parts of the world over the last century.

Our focus is on the Spanish Inner Colonization, a large-scale agricultural and land settlement program executed during the Francoist dictatorship, with a special interest in the use of technologies of visual media. We will show how this program was understood as a process in which land was partitioned and monitored with image technologies, and involved these operations simultaneously on multiple scales. First, the program developed plans where vast extensions of land were designated as irrigation zones; then, to be operative, these zones were segmented into rational agricultural units or plots; inside them, finally, the plots were gradually administered as technical surfaces quantitatively monitored. We will emphasize how in all these scales, the role of the image was constitutive, and, in addition, we will discuss how these operations were part of the industrialization of agriculture that took place in the twentieth century and how they also facilitated the evolution of agriculture to become, literally, a visual practice performed by mechanical devices operating in the air and on the ground. While it might seem technologically more rudimentary than some of the more recent precision agriculture schemes that employ drones, cloud technologies, and vast datafication of crop fields, they are part of a genealogy of technics of agriculture that is also about techniques of images. Aligning with Ross Exo Adams's landscapes of posthistory, we want to consider how these programs evidenced the idea of programmability of agricultural landscapes as a project that incorporated different political, ideological, and historical layers as to how living surfaces became infrastructure.[13]

SPANISH INNER COLONIZATION

At the end of the Spanish Civil War in 1939, the country faced devastation in several territories exposed to a three-year period of intense destructive forces. The resulting military government, the Francoist dictatorship, launched several postwar reconstruction programs aimed mainly at restoring the destroyed infrastructure and geared toward attenuating the starvation effects of the brutal war among the population. In parallel, other plans were also quickly initiated and designed to support the recovery of economic activities, such as the development of industry, housing, and agriculture. These were long-term plans managed, respectively, by the Instituto Nacional de Industria (industry), Instituto Nacional de Vivienda (housing), and the Instituto Nacional de Colonización (agriculture; from here on abbreviated as INC).

We will examine the operations carried out by the agricultural institute, a colonization plan that involved the expropriation of plots, the introduction of new techniques—water infrastructures in particular—and the accommodation of new settlers. Called Inner Colonization (*colonización interior*),[14] the aim was to augment the production of crops in order to align them to the needs of national demand and, afterward, for export. Furthermore, the proposed rationalization of space in terms of units of agricultural practice and the subsequent transformation of the economic and social structure were developments whose interweaving with visual operations has not so far been fully addressed.

Many similar plans, albeit with local differences, had already been tested in other countries. Most of them took place before the Second World War, a fact that highlights that the Spanish plan was a late development. In an exhaustive study carried out by the Spanish government in 1925, thirty-four countries were analyzed in relation to their existing agrarian and repopulation reforms. *La colonización y repoblación interior en los principales países y en España* (The colonization and internal repopulation in the main countries and in Spain) (1925) published during the dictatorship of Miguel Primo de Rivera (1923–1930), consisted of three volumes, in which numerous cases were analyzed, including those of Algeria, Argentina, Australia, Austria, Belgium, Brazil, Canada, Denmark, Egypt, England, Finland, France, Germany, Hungary, India, Ireland, Italy, Japan, Korea, Mexico, Netherlands, New Zealand, Norway, Portugal, Russia, Siberia, South Africa, Sweden, Switzerland, Tripoli, Tunisia, United States, and Uruguay.[15]

The common dimension of many of the projects concerned an intensification of agricultural production and the settlement of new population; these two aspects also identified them as colonization processes, specifying in some cases that it was an "inner" or "internal" colonization, in contrast to the more usual outward sense of the word.[16] Some of them are remarkable examples as they inspired, in turn, the Spanish one: Mussolini's several *bonifiche agrarie*,[17] the Portuguese Internal Colonization (*colonização interna*), the German *innere Kolonisation*, and the Israeli moshavim and kibbutzim.

For example, the German case of agricultural economist Max Sering illuminates well the historical continuity of imperialism with agricultural projects. What ended up as part of the national socialist invasion of Poland as part of the broader "inner colonization" of the East had its earlier roots in Prussian agricultural planning. Sering's 1883 trip to Canada was part of the broader scouting for methods that would become part of the European plan to implement settler colonialism toward the east. "North America's Wild West would be translated into Germany's Wild East" sounded the ambitious task that aimed to provide a design template for a reformatting of vast geographical areas and landscapes.[18]

Indeed, these comparative observations were premised on and repeated the mantra of *terra nullius* as the basis for redesigning "unused" into "productive" land: "From the metropolis of Winnipeg and westward towards the frontier, he saw a government-sponsored, seemingly rational allocation of land to settlers—settlers who then ventured forth, ordering the wild, 'empty' landscape, colonizing the land and fundamentally strengthening the Canadian government and the Canadian race."[19] The settler colonial tropes were the discursive frame packed as part of a plan to "to send wagons east and Germanize the land" in late nineteenth century Europe, too.[20] The trope of *Lebensraum* was already part of the pre-Nazi period, but the historical premise of this space was not just one for humans, but for "the pigs, the potatoes, and the wheat [that] embodied fascism,"[21] and also the administrative policy plans of inner colonialism that had existed since the nineteenth century, before the national socialist period.

In addition, besides the international examples, which in some cases became inspirational guidelines, it is important to recall that this inward mode of colonization had been practiced in the context of Spain long before Franco's dictatorship. Since the eighteenth century, and with particular emphasis during the second half of the nineteenth century, the modernization of agriculture involved law-enforced expropriation of terrains.[22] Many territories were ruled by powerful landowners, so expropriation was the only means to enforce, on the one hand, a greater distribution of land property and, on the other hand, to develop irrigation infrastructures. So-called colonizations had been put into practice since the eighteenth century and were conceived as a politics of rationalization of land use, combining the foundation of settlements with modernized agricultural methods.[23]

What distinguished the colonization by the INC in relation to these precedents was its scale. Undertaken by an authoritarian dictatorship over three decades—from 1939 to 1973—the actions of the INC involved the excavation of a vast network of water canals, the setup of large irrigation zones around the main river basins, and the foundation of approximately 300 new settler towns to put them into production (figure 4.1).[24] It merged three main strategic lines as a program: an agrarian reform, a land settlement program, and a hydraulic project.[25] And, singularly, only one institution managed it, the INC, which concentrated the power to design and execute plans that involved an immense variety of procedures: expropriations all over the territory, large movements of population, infrastructure works, drainage of ponds, soil leveling, the supply of machines and fertilizers, formative practices, finance capabilities, and centralized management of the information gathered in a continuous monitoring of the process.[26] It was, therefore, a huge institution at the time, and it gave rise to one of the most ambitious reforms in the recent economic history of Spain, "the largest urban operation within rural zones ever practiced in Spain,"[27] which contributed to changing, as a result, the face of its rural landscape. Management of information and management of soil were closely related and, as we will argue, both mediated by the presence of operational images of multiple kinds.

When seen retrospectively, the temporal and spatial extent of the inner colonization favored the spread of irrigation infrastructures and the introduction of new techniques within Spanish agriculture.[28] In fact, it resulted in doubling the overall irrigation surface while refashioning the map of cultivation and exploitation (figure 4.2).[29] As a colonization program, however, it failed to redistribute land as property due to several loopholes and practices of political clientelism.[30] It operated, instead, as an ideological tool of a paternalist state, an overscaled investment that was nevertheless profitable from the point of view of the political propaganda of an authoritarian dictatorship. Despite its economic inefficiencies, it

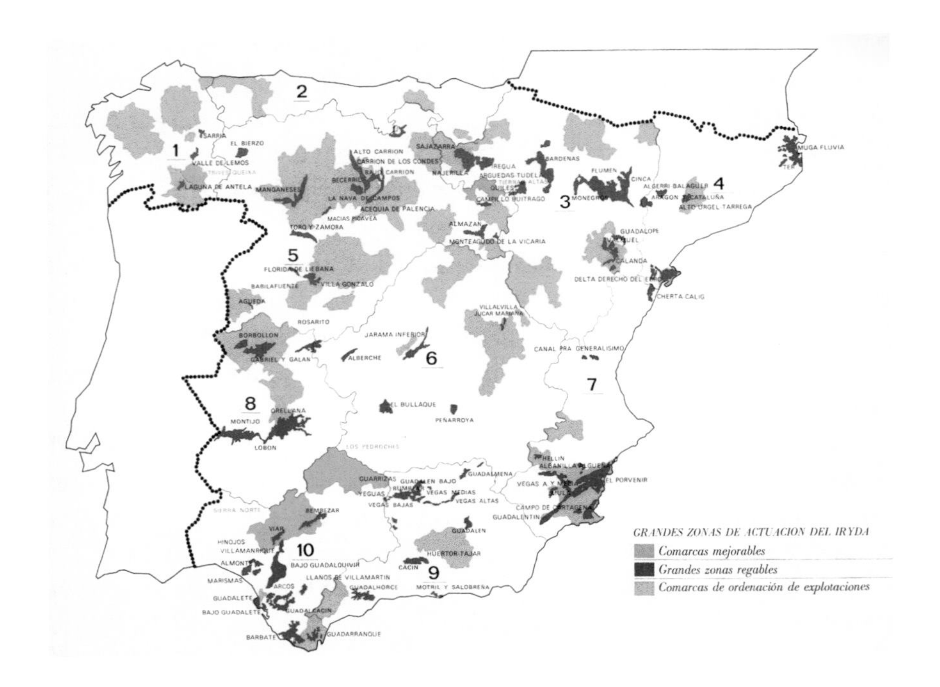

4.1

Map of the irrigation zones (green) developed by the Instituto Nacional de Colonización (INC) and by the Instituto Nacional de Reforma y Desarrollo Agrario (IRYDA), which continued its development activities (yellow and pink) as part of the democratic state. Source: "Grandes zonas de actuación del IRYDA," in *IRYDA: Fines y actividades* (Madrid: Ministerio de Agricultura, IRYDA, 1973), 5. Copy and image from the collection of the Biblioteca del Ministerio de Agricultura, Pesca y Alimentación, registration number 3719. Used with permission.

4.2

Illustration of the irrigation zone of Las Bardenas as a green mantle. Produced by the IRYDA, the institution of the democratic state that took the place of the colonization program. Source: "Zonas Regables. Las Bardenas. Aragón," in *IRYDA: Fines y actividades* (Madrid: Ministerio de Agricultura. IRYDA, 1973), 7, and transparency. Copy and image from the collection of the Biblioteca del Ministerio de Agricultura, Pesca y Alimentación, registration number 3719. Used with permission.

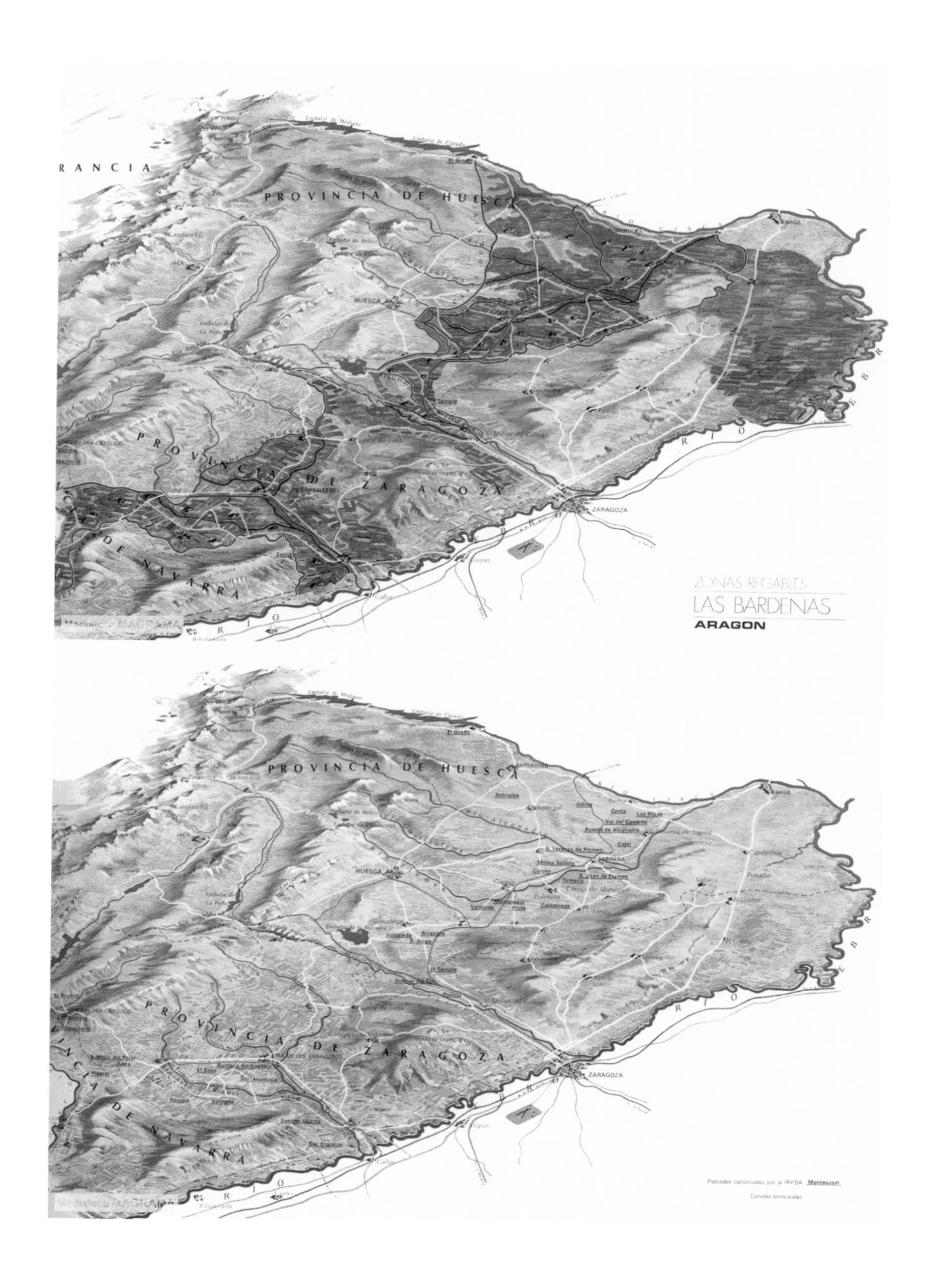

RANCIA
PROVINCIA DE HUESCA
HUESCA
PROVINCIA DE ZARAGOZA
ZARAGOZA
PROVINCIA DE NAVARRA
RIO EBRO
ZONAS REGABLES
LAS BARDENAS
ARAGON
Poblados construidos por el IRYDA Montesusin
Canales principales

still contributed to the extractive relationship to soil: the gradual transformation of agriculture into the fully mechanized and industrialized land practice would turn some areas of Spain, despite their low precipitation levels, into major exporters of fruit and vegetables.

AN IMAGE-BASED COLONIZATION

Several studies have attempted to provide a complete picture of this vast and violent colonization program. These studies include detailed documentation about the evolution of the program, its extractivist and repressive nature, the use of prisoners as slave labor, the relationship with previous colonization programs and with similar international developments, and its evolution and impact decades later.[31] Despite this wealth of research, and with the exception of one volume dedicated to the photographic legacy that resulted from the documentation and propaganda practices of the process,[32] the importance of the role of images in the design and execution of the program has not been sufficiently analyzed. Certainly, the areas affected by the colonization schemes underwent profound changes in their landscape. However, to what extent did the image play a role in this process? The language of the head of one of the delegations of the Instituto Nacional de Colonización points to a scheme of operations rooted in the image: "In these years, in those inhospitable places, beautiful and clean villages have sprung up, thousands of trees have grown, the bare white lands have been painted with the splendid green of the alfalfa groves in the backlight of the sunset; aridity has been succeeded by lushness; despair by hope; resentment by moral beauty."[33]

The quotation highlights the redemptive character that the regime attributed to this and other technoscientific programs.[34] Landscapes are described through immaterial sentiments; programming land is programming affect. However, the phrasing "the white lands have been painted green," used to describe the colonizing work, indicates the approach to the landscape that we want to analyze. Implicit in the description of the existing lands as "bare white," we can read the ideology of abstraction and invisibility characteristic of colonial territoriality—as described, for example, by Brenna Bhandar—which has been explicitly traced in other agricultural colonization programs too.[35] But it is the phrase "the bare white lands have been painted with the splendid green of the alfalfa groves"—its distance to the territory and the emphasis on the visual transformation of the landscape—that deserves more detailed attention so as to carve out what it means specifically in this historical context and more broadly, in the context of our book.

The quote by Francisco de los Ríos Romero describes the result of a process that was already in its final phase, with most of the main plans already executed. From two decades or so earlier, in the same year that the law was passed that gave way to the most expansive phase of the program,[36] came a documentary by the Marqués de Villa Alcázar, commissioned by the Ministerio de Agricultura, Pesca y Alimentación, *España se prepara* (1949), which contains a series of sequences that illustrate the same idea of the institute's delegate, although in this second case in cinematographic terms. At one point in the documentary, a succession of shots displays various characteristic elements of the technoscientific material culture related to the colonizing program: dams, instruments for chemical analysis, scenes of agronomic training, and accounting machines, and so on. The sequence ends with the image of a technician from the institute opening what appears to be one of the colonization guideline books. On its pages, a sketch of a street view of one of the colonization villages is shown to the camera (figure 4.3). Then, the drawing is blended with the moving image of the already-built village. The notion of a projected space that is born on paper and ends up being implanted in a territory is explicitly staged.[37] In this regard, the idea that the colonization process has given rise to a series of villages that materialized a mix of avant-garde design, rurality, and production of and through images is pervasive: "The aerial view of the new settlements, as shown in photographs of the period, displays them as villages that had been built to be photographed from the air. Their construction, starting from scratch, facilitated this design. They appeared to be toy towns. This facilitated the tendency to window-dressing that remained within the mentality of the Regime."[38]

Moreover, the documentary continues with the visual trick of making the image of the transformed lands emerge out of drawings. The sequence

that follows the materialization of the village repeats the same visual effect but with different elements involved. A sketch illustrates this time a crossroads on a large esplanade where two vehicles meet (figure 4.3). The drawing is then blended with the moving image of the crossroads. It is the moment when two cars meet; one of them slows down to give way to the other. The image here materializes not a stabilized element of architecture but an anecdotal event that occurs in the transformed territory. This time we are not in front of a plan or a drawing projected toward what has been built but of a sequence that evokes an ability to foresee and manage the events that are about to take place. This second series of images

4.3

Sequences from the documentary film *España se prepara*, directed by Jesús Francisco González de la Riva y Vidiella, Marqués de Villa Alcázar, 1949. Public domain, stills by the authors. https://www.youtube.com/watch?v=WAG2FrKzeRI.

recalls the idea of a designed future. It speaks for a domain of transformations from the scale of the landscape to the individual events of every day, where not only space is partitioned, but time is regulated anew.

This idea has already been proposed in relation to the photographic documentation of everyday life during the first years of the Spanish inner colonization. In these archives, we see images of displaced settler families—with their past left behind—against the background of newly built villages. Their rational and serial architecture projects them toward the future. The contrast between the past left behind by the settlers and the setup of a future already designed for them is remarkable.[39] Not by chance, these images of crossroads and architecture were taken in a territory whose sowings and harvests were going to be regulated and monitored by the centralized INC. This same institution would oversee and administer the details of the settler families' daily lives and their performance in the countryside.[40] In this regard, the visual territoriality of this colonization is a space that needs to be examined: how the agroindustrial regulation of vegetal life through images entailed a model of control of all the everyday spaces of life, in line incidentally with the needs of the repressive dictatorship. To examine this, we will switch to the context of the technical images mobilized during the design of the programs. Their dominant role will become evident.

The Seen and the Seeded

Cultural techniques such as the grid are practiced, learned, and disseminated in time and space, and they are also embedded, reshaped, or codified with different tools, devices, or media. Grids function as operative ontologies that do not merely represent but materially transform territories and their affordances, that is, their enabling capacities.[41] This is the case, for instance, with photogrammetric equipment used in aerial surveys. Equipped with GPS receivers and other movement-tracking devices, digital cameras for aerial photography can produce images suitable for being automatically rectified with appropriate software and exported to fit on grid-based tiled maps.[42] Although the first military and commercial procedures of aerial photographic surveys needed a lot of time and the labor of skilled interpreters to build up photomosaics with the aid of existing maps, they have been used since the 1910s systematically as a measuring tool.[43] Through them, military and civilian infrastructures have been located and cadastral information has been retrieved and added with precision to the grids of maps. Aerial surveys became particularly apt for twentieth-century development and land reform plans all around the world, where the scale of operations, such as water infrastructures or urbanizations, meet the spatial extent of the aerial perspective.

In 1926, during the dictatorship of Primo de Rivera, a new administrative entity was established in Spain: the Confederaciones Hidrográficas (River Basin Authorities). Instead of the province or other political-geographical demarcations, the physical river basin and its connected waters became the territorial unit used to integrate water planning.[44] In this way, one single institution could supervise all possible uses of fluvial waters, such as irrigation, transport, and energy. The law of 1926 that formalized the creation of the basin authorities also demanded the complete and precise cartographies of the territories under their control, detailed enough to display their divisions into plots. As existing resources were inadequate, and in order to acquire this material quickly, the services offered by a private company that pioneered and promoted aerial photography were contracted as the only means to render it technically possible.[45] The first set of official aerial photographic images of Spanish land date from this time, and the use of aircraft to produce cartographic documents was soon extended to the completion of an updated cadaster. The number of acres of land photographed from the sky then grew—until the Civil War brought everything to a halt.

Years later, after the Second World War, the US Army Map Service carried on its own aerial mapping project, completing two orthophotography surveys of the whole Spanish territory, in 1945–1946 and 1956–1957.[46] They belonged to the so-called Casey Jones Project, devised to map from the air the territories of Portugal, Italy, and Spain, as part of a defensive strategy against a hypothetical conflict with the USSR.[47] Remarkably, although the resulting maps were kept secret and not shared by the United States with the Spanish Army until 1959, copies of the first flights in the 1940s have been found in one—and only one—of the archives of a Francoist institution: inside the INC, the institute responsible for the irrigation programs.[48] Considering the initial secret character of these images, this should be considered proof of the INC's strong interest in the aerial photography.

Two intersecting cultural techniques were at play, and both concerned surfaces. During the Spanish inner colonization, the same land was measured, parceled, and populated, as well as being photographed frame by frame by fleets of aircraft. In some respects, these were two coupled and connected envelopes growing at the same time: a surface of hundreds of thousands of acres of uncultivated land transformed into green areas of productive yields, linked in some direct and indirect ways with the organized grid of images taken from airplanes, organized into ever-larger arrays of photographic and map surfaces. Figure 4.4 shows the evolution of these extensions plotted over the basin of the Ebro River. It overlaps three moments of this evolution. The light gray areas display the irrigation zones that were colonized in plans before the Francoist program, during the second half of the 1920s. The grid of aerial photographs linked to the law of 1926, commissioned by the River Basin Authorities to the private company CEFTA in 1929.[49] Finally, the dark gray ones, drawn up fifty years later during the 1970s, depict all irrigation zones produced by the dictatorship's plans within this basin. When overlapped, these images coincide in striking ways: the gridded areas mapped during the first cartography of the Ebro basin overlap with the spread of the dark gray irrigation zones afterward, during the Francoist colonization. Almost every zone on the left bank of the River Ebrolies within the areas mapped during those first flights.

The images of the Ebro basin acknowledge the principal role that aerial vision had in the design of these transformations of the territory. This feature is also emphasized by the exceptional presence of copies of photographs taken on US flights in the archives of the INC. Seen from this sequence of layers, the squares of the aerial grid seem to extend the initially colonized terrains, depicted in light gray on the map. The new irrigation zones created by the INC, the dark gray areas, thus seem to appear as a material development of the aerial photographs. To observe them as temporal cuts in a fifty-year line suggests how the agriculture practiced in this colonization was akin to a photographic transfer: it is as if the vegetables had been printed on the terrain after the image. In other words, displaying the large-scale practice of agriculture that took place in Spanish colonization plans against the extent and spread of aerial images allows us to understand these agricultural operations not only in relation

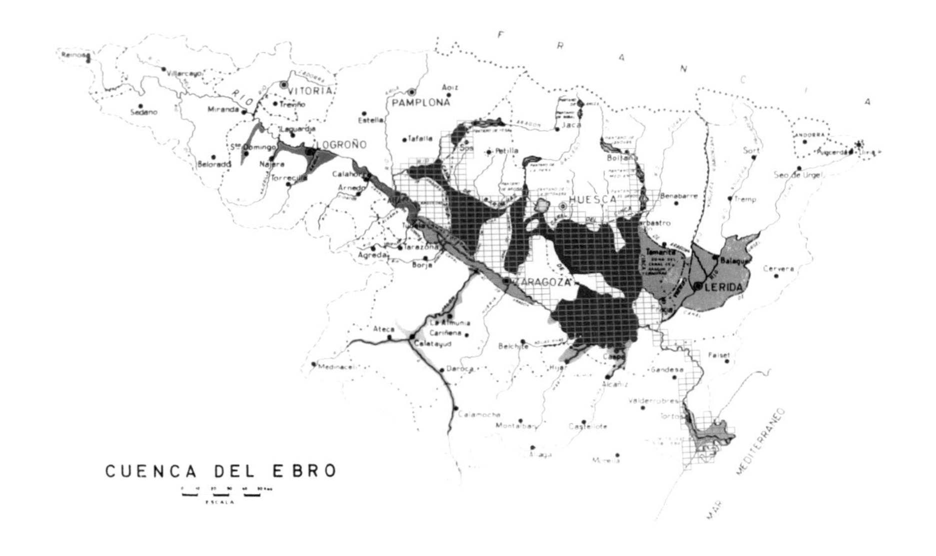

4.4

Maps of the Ebro River Basin, with colonization zones before Franco (light gray), the areas covered by the first aerial mapping (grids), and the zones by the INC (dark gray). Elaborated by the authors from data from Felipe Fernández García, "Fotografía aérea histórica e historia de la fotografía aérea en España," *Ería* 98 (2015): 217–240, at 224. Source of the map: "Zona de actuación de la Delegación del Ebro," Ministerio de Agricultura, Pesca y Alimentación Archivo Central, Fondo XXV Aniversario Instituto Nacional de Colonización–Delegación del Ebro-Teruel, signature 03 Mapa Cuenca del Ebro. Used with permission.

to the photosynthetic activity of plants but also in relation to the photosensitive surfaces of the photographic plates.

The use of aerial photographs starts to suggest the emergence of a visual mode of cultivation—literally, a visual agriculture that concerns the modulation, control, and distribution of light as plant growth or as (operational) images. The inner colonization did not seek to expand the surfaces of arable land through outward colonizing practices but rather to intensify agricultural production by modernizing it. This new agriculture needed to establish a new relation with soil.

Soil alone, an *idle soil*, became a parasite inside the domains of a state that looked for productivity. As one of the members of the INC wrote, "At the sight of so many lands that will be fertile, it seemed to me that the whole land of Bardenas was a sleeping paradise."[50] The soil needed to be mapped, conveniently recognized, and identified in order to be awoken and put into production. With the aid of the addressed space of the photographic grids and with the measuring and monitoring details of aerial photogrammetry, agriculture became a media-based activity that drew on several cultural techniques. The parasite needed to be controlled; it became, in Michel Serres's terms, the excluded third and gave rise to the return to a media-driven definition, apprehension, and operationalization of soil and agricultural production.[51]

The Visual Cultures of Precision Farming

Thus far, we have located the role and scope of images and aerial photogrammetry in the design and communication of the Spanish inner colonization. This program is a significant case study in its own right, but also for how it taps into a particular kind of process of agriculturally driven colonization that had featured in different forms in different geographical areas, including in German plans for expansion toward the East (modeled after North American settler colonialism) and in how the Russian Empire related to the vast geographical areas under its control.[52]

Before continuing with more details of the Spanish case, it is worth observing that the development of visual-based soil operations has evolved to become a multiscale practice today in a much more dense and intensive way. Think of the following scene as a diagram of events. A ray of sunlight arrives on the earth and meets the leaf of one of the vegetables produced in these zones. Photosynthesis takes place, and a ray of green light—with a bit of red, too—is reflected to the sky, where it is trapped by a photographic plate and carried by an aircraft, which is part of a program that is transforming the terrain where the plant is being grown. This diagram, which seldom occurred during the aerially enforced colonization, grounds the practices of contemporary agriculture. Radiation of light bounces off surfaces, it is captured, and as captured data and images, it is bounced back to become operationalized landscapes of growth.[53]

Since the 1990s, thanks to the availability of satellite imagery, the umbrella terms *precision agriculture* and *precision farming* refer to the use of remote sensing in agricultural contexts. They embrace an expanded set of practices where aerial vehicles track and guide the actions executed on the ground. For instance, devices on tractors are programmed to control the dispersion of water and chemicals at each point of the plot from information gained from aircraft-based or satellite sensors that measure the wavelengths of radiant energy absorbed and reflected from the surface of the land. The infrared light reflected by the earth informs the satellite of its biological activity. When read against averaged models and statistics, this information can be used to detect whether a particular plot of soil needs water, fertilizers, or pesticides. The resulting diagnosis, obtained from the air, is sent by sensing devices—drones, airplanes, or satellites—back to the ground as images called *prescription maps* (figure 4.5). Tractors provided with so-called variable rate applicators—digitally controlled irrigators—receive these prescriptive images. Soil moisture, surface temperature, photosynthetic activity, and weed or pest infestations are addressable with a resolution smaller than a square meter, almost exactly the size of the irrigation system actuator. Such a circulation of data and images becomes part of the material flows as part of the management approach. Within this circuit, the ground becomes a screen, read as a stream of images that specify the doses to be applied at each point, with tractors modifying the states of each pixel.[54]

At a different scale, the Spanish inner colonization is part of this media genealogy of cultural techniques. As a large-scale program that needed to recognize the unproductive lands to be transformed—identifying existing

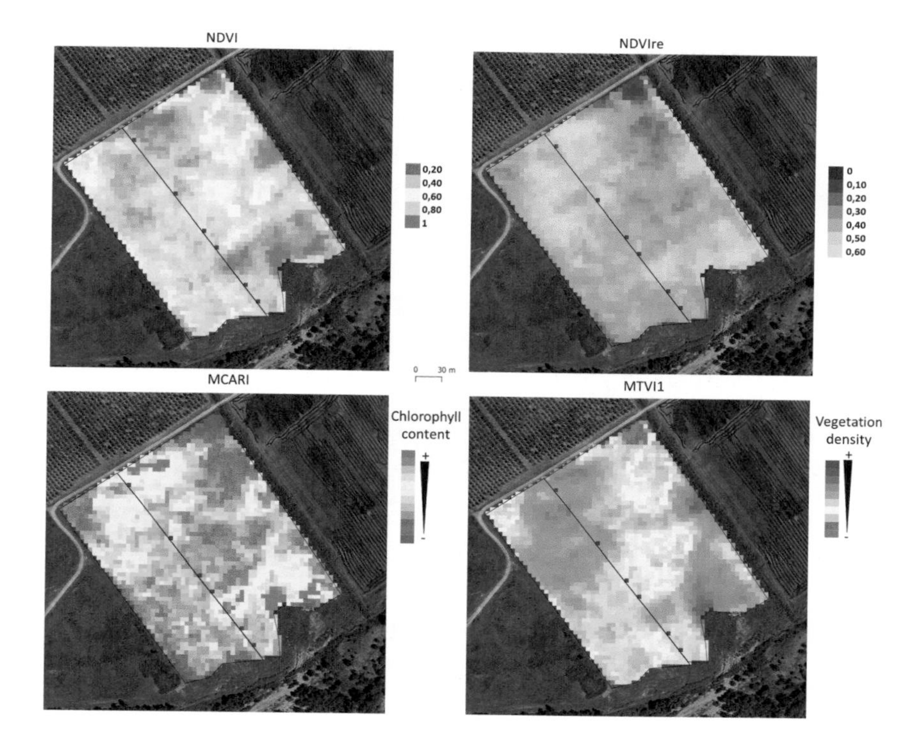

4.5

Precision agriculture maps of multispectral indices obtained from elaboration of satellite data related to wheat cultivation. Source: Tiziana Simoniello, Rosa Coluzzi, Mariagrazia D'Emilio, Vito Imbrenda, Luca Salvati, Rosa Sinisi, and Vito Summa, "Going Conservative or Conventional? Investigating Farm Management Strategies in between Economic and Environmental Sustainability in Southern Italy," *Agronomy* 12, no. 3 (2022): 597. Used with permission.

plots, evaluating their character and crop figures, and addressing the owners to be expropriated—aerial imagery and photogrammetry became prescription tools to shape the selected zones. These, in turn, became the subject of the operations of a centralized unit, the INC, which mobilized the tractors, excavated the water streams, and set up the workforce. The chain of operations spread from airplanes to images to water distribution and organized labor. As in the case of precision farming, it was a visual circuit meant to activate unproductive soil to populate seemingly empty lands. But in this earlier context, the bureaucratic paper machine was essential to establishing this information-data-agriculture logistical circuit on an institutional level instead of merely seeing this landscape as *technological* programmability.

INFRASTRUCTURAL SPACE

As briefly mentioned, the case of the Spanish inner colonization we are analyzing presents already some of the key elements of precision farming. One is the circulation between soil and aerial images, and back to plants and plant growth, even if the images in the later practices are quite different, acquired through advanced remote sensing techniques and processed as data analytics.

Much of our interest focuses on the logistics enabling this circulation. In chapter 1, we showed how a material culture of glass containers gave rise to a model of the plant-environment relation, as well as to an unprecedented global circulation of plants. Before the photometric models of the plant discussed in chapter 2, plants had already been abstracted and circulated into an infrastructural space of logistics. Similarly, in the case of the INC, its actions entailed an approach to space and time which was fundamentally logistical too. In this regard, INC started to generate agricultural landscapes as sequences of easily recognizable patterns and architectures, such as the constellations of small towns, branched irrigation channels, and slim forest outlines to surround the transformed zones. Through this "soupy matrix of details and repeatable formulas,"[55] the infrastructure space of the zones was deployed as "spatial software,"[56] the very guidelines of the colonizing program.

A singular component of these guidelines was the protocol used to address the positions of the newly built towns. As a state-driven program, most of the relevant actions of the inner colonization plans were logged and published in the *Boletín Oficial del Estado* (official state gazette, *BOE*). There, remarkably, we find the announcements of new settlement towns, together with their proposed locations. In order to address their positions, a system based on the road network was used. From 1940, roads and motorways were numbered following a procedural nomenclature system.[57] This organized platform of road names served as a reference for the position of the new towns to be created inside the lands to be colonized. Thus, their locations were written down as distances to the nearest roads, specifying the relevant kilometric milestones.[58] The location of Pizarro, Hernán Cortés, and Alonso de Ojeda, for instance—tiny towns significantly named after the *conquistadores* sent to America[59]—were specified in the following way (as shown in table 4.1):

As an address system based on the calculation of distances to infrastructure in the background, this information evidences the operational transformation of territories after the impact of transport infrastructures. Here, as in the road infrastructure spaces analyzed by Harvey and Knox, geographical surfaces are conceived and designed after a new kind of spatial awareness.[60] Space is apprehended in relation to the effects of a calculating background, a logistical mechanism that "activates" space. As Nigel Thrift has explained, it is a "movement-space" that is linked to the experience of the gridding of time and space, the invention of filling and listing systems, and the invention of logistics.[61] Surfaces of land are

Table 4.1
Locations of new settler towns as specified in the *Boletín Oficial del Estado*, July 5, 1955

Pizarro	Subzone C, within the boundaries of the municipality of El Campo, 5 kilometers south of this town, counted on the road L-420, and 300 meters west of this road
Hernán Cortés	Subzone E, within the boundaries of the municipality of Don Benito, to the north and next to the turnoff of road N-430, in the stretch between Santa Amalia and Valdivia, next to the bridge that will be built over the Ruecas river
Alonso de Ojeda	Subzone G, within the boundaries of the municipality of Don Benito, in the vicinities of the crossing of the road from Cuadrado de Almoharín to Santa Amalia with the borderline between the provinces of Cáceres and Badajoz

intertwined within a new spatiotemporal continuity that emerges in the practices of sorting, numbering, and calculating, where "it is relationality that is important . . . turning space and time from 'a priori' into 'a posteriori' categories."[62] The nodes of the networks of settlements were thus announced in the official gazette through messages that evidenced the projective logic that reshaped the colonized territories.[63]

Interestingly, the logic of circles and radial distances that feature prominently in the centralized logistical space of the road system also characterized the partitioning of the space inside the zones and gave rise to some of the most remarkable cases of urban design in the newly built villages. The soil that was going to be photographed from the air was in parallel being mapped by way of logistical media on the ground, through the auxiliary partitioning by way of transport infrastructure. Echoing John Durham Peters's words: "The job of logistical media is to organize and orient, to arrange people and property, often into grids."[64] In the case of the Spanish inner colonization, one can observe at different scales the pervasive presence of the circular partition of space, a fundamental cultural technique in its own right that establishes hierarchies of governance, property, and control. As we will see next, circular arrangements modeled after techniques of drawing and inscription were used to subsume the material flows that produced the agricultural crops: workers and water. Their labor, according to Rossiter's work on logistical media, "underpins the traffic of infrastructure and circuits of capital."[65] From the point of view of the circuits they belonged to, workers and water operated as components of the new synthetic environments.

The Cart-Module: The Footprint of Visual Agriculture

Early large-scale agriculture entailed the expertise of geometry. Where the flooding of the Nile erased the parceled banks, the land surveyors—the *harpedonapts* or rope stretchers—with their ability to calculate areas from their measurements, were the ones who could bring back the agricultural order in the midst of shifting landscapes. In Michel Serres's words, "They had the cord, the unit, the measure, writing, and prestige."[66] The equilibrium between landowners was articulated by the size of the plots, which came out as the critical magnitude.

In the case of the agricultural program we are focusing on, property and size were not the problem. First, expropriated terrains were considered to be lands of national interest that belonged to the state. Second, following the procedures put into work in the referential case of the US Columbia Basin Project, the shape of the family units and the sizes of the plots were prescribed as part of the colonization plans.[67] In other words, the zones were partitioned in a homogeneous way. The main problem of the plans, however, resided in the spatial distribution of the houses of the settlers and the allocation of their plots within the irrigation zones, where nothing existed but wasteland or previous dry exploitations. Considering the productive spirit of the whole project, the INC needed to find a spatial distribution that filled the entire available space with agrarian units—house plus plot—in such a way that no idle soil could be found within the zones.

After analyzing the failure of Mussolini's model based on disseminated farms,[68] the INC decided to distribute settlers among constellations of towns scattered as networks inside the large irrigation zones. The problem was transformed then into a question about the spatial distribution of the nodes of a network. Additionally, each constellation of settlements had to be linked to the irrigation system, as the settlers needed to be placed in the vicinity of their irrigated plots. This problem was solved with one of the most salient features of the Spanish plans: the use of the so-called *cart-module*. The cart-module was a graphic planning tool used by INC that was defined by the maximum operative distance covered by a settler with a cart; that is, the distance that would allow a farmer to go and return to a plot without losing much time.[69] This distance, estimated as 2.5 kilometers, was used as the radius of a circle around the town—the cart-module, or its area of spatial influence—where plots were placed. Neighboring towns, therefore, should ideally be separated by 5 kilometers, with their areas of influence drawn as tangent circles. The cart-module consequently allowed the graphic exploration of different combinations able to fill the available space. Once one of these visual arrangements was chosen, the circles were transformed into the canals, the water, and the humans that would farm the zones (figure 4.6).

This drawing technique allowed INC technicians to position the settler towns and deploy the needed infrastructure of water canals. In

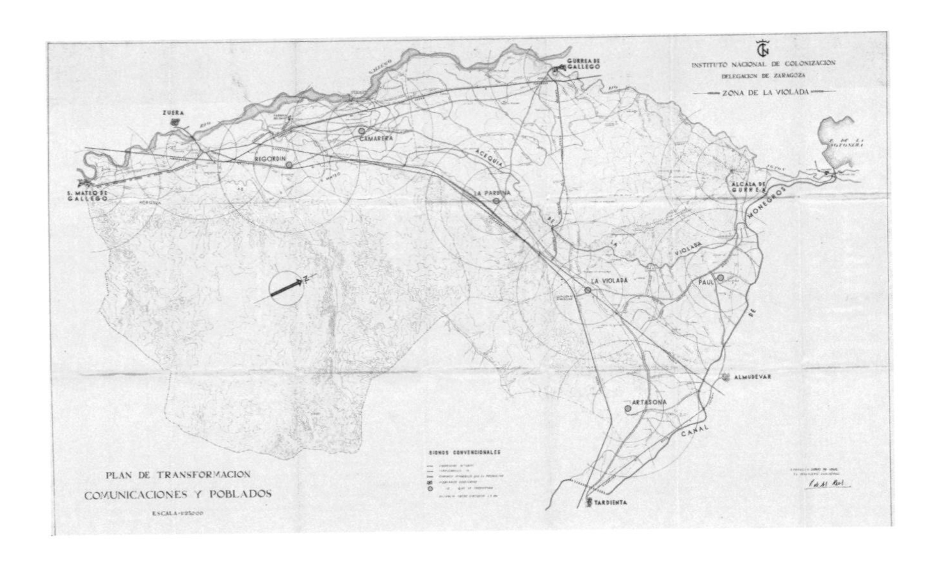

a singular case of mapping infrastructures, the central role of the cart-module highlights a significant dimension of agriculture put into play: agriculture was being, in Lisa Parks's term, broadcast over these prescribed distances. In Parks's work, broadcasting is a technologized practice that results in the establishment of signal territories or, in other words, territories that are both culturally and materially transformed by the presence of specific signals.[70] Within the inner colonization, the design of the

4.6

Cart-modules in the irrigation zone of La Violada. Source: "Plano no 8 del Proyecto General de Colonización de la zona declarada de interés nacional, dominada por la acequia de la Violada (Huesca-Zaragoza). Planos sector III [España] (1943)," Ministerio de Agricultura, Pesca y Alimentación, Archivo Central, Fondo Archivo Central. Proyectos Instituto Nacional de Colonización–INC, Proyecto no 148. Used with permission.

cart-modules abstracted the daily flows of farmers and water streams as periodic signals whose extent was calculated beforehand. On the maps of the infrastructure, workers and water are replaced by circular signals broadcast by the settlement nodes of the network. The zones became signal territories, where each single settler town assumed its own footprint whose territorial boundary was defined by the cart-module.

The arable space emerged after the expropriations as a signal space. In addition to the space-defining character of the cart-module circles as particular signal territories, it is important to highlight their time-based nature. As we have explained, their radius was linked to the workers' displacement time considered in relation to their workday. It was an averaged everyday measurement of human activity that shaped the vast zones. If water had to reach the plots through the network circuits, farmers had to access them by foot or with the aid of animals.[71] The ordering of space was subsumed to this temporal constraint, which was furthermore related to labor time. Hence, the cart-module was a cultural technique working simultaneously on the laboring body, the organized plot, and the temporal coordinates on which these different elements were synchronized. It established then a set of conditions similar to the famous example of plowing as a cultural technique that Cornelia Vismann has written about. Such techniques are self-referential, as their operation establishes their fundamental elements of reference from the material line on the ground to the philosophical and juridical establishment of a subject as *the effect* of the act of drawing a line. Furthermore, when Vismann refers to the spatial techniques (borders, surveying) and to the genealogical techniques that "govern notions of duration, assign origins and secure the future [such as] record-keeping, adoption and inheritance regulations, but also breeding and grafting,"[72] we can add that many of the agricultural media techniques do both: they work on the land while also working on the different temporal regimes where growth but also labor time becomes managed.

Photographic Environments

Together with the images of dynamically ordered landscapes and infrastructures, the emphasis on building a new world is particularly striking in the new towns erected to host settlers. Today, these urban experiments are well known because of their intrinsic architectural qualities. After the failure of a model based on disseminated farms, settlers' towns were created, inspired by community-centric urban models, such as the Neighborhood Unit or Radburn planning. Notably, their rational and serialized architecture, as well as their overdesigned urbanism, favored aerial shots. The settlements, seen from the air, appear as scaled mock-ups, freshly emerged from the drawings, in a state still between the graphic and the physical, echoing in some of its blueprints the circular pattern that pervaded the colonization (figure 4.7).

The abstract and serialized urbanism did not mitigate the settlers' sense of loss when inhabiting the new villages. They did not own houses and plots: moreover, for twenty to forty years, they had to pay off the housing and investments provided by the institute (which was, meanwhile, the legal owner). Instead, theirs was an environment turned into a production system that was also used as a propaganda tool for the dictatorship. Consequently, their everyday life took place in an utterly controlled environment. Dwellers were individually selected, trained, and monitored to avoid problems inside new towns. In addition, their continuity in the program relied on a contract that entailed many forms of monitoring. This was a bureaucracy that managed, at the same time, domestic issues, the maintenance of towns, and the productivity of fields. It also included up to ten different kinds of guards and keepers, who were dependent on a public administration and the power of a public servant or inspector working for the INC, who could expel any settler if he or she did not "meet with their obligations normally."[73]

To keep the big land tenants happy, most of the plots offered to settlers had low-quality soils. Also, numerous settlers had never worked before with irrigation techniques.[74] Initially, during the program's first decade, these limitations resulted in very low yields. Only after a redesign of the guidelines did each of the settler towns receive an agricultural engineer from the INC, with each of the zones under the responsibility of a chief engineer. Thus, INC assumed control of every aspect of everyday irrigation: its design, the types of cultivation, timings, fertilizers, and pesticides. The inner colonization became a program of supervised agriculture, with settlers receiving specific training and being monitored over a two-year testing period. Additionally, some of the urban plans

emphasized an interweaving between population and soil: in Cañada del Agra, a root-like spreading of the streets allowed the town to organically lay on the terrain; in many other towns, such as Esquivel (figure 4.7), the main square was placed outside, as if crops were incorporated into the urban scene. It is as if settlers and crops were not necessarily distinguishable, as if an underlying managerial and logistic space were addressing and symbolically manipulating settlers and crops, as if they had already

4.7

Settlement town of Esquivel (Seville), built between 1953 and 1955 as part of the inner colonization. Source: "Esquivel, Alcalá del Río (Sevilla)," Ministerio de Agricultura, Pesca y Alimentación, Archivo Central, Fondo INC Instituto Nacional de Colonización (1940–1970), signature 11033-c14-cd1-esquivel. Used with permission.

been transformed into data. The irrigation machine, then, was put into operation.

The industrial nature of the colonizing agriculture analyzed here involves a controlled cut in the time and space of the sedimentation of light on infrastructured surfaces of soil. Light is commodified by means of the spatial control of flows, such as water and pesticides, through gates and exposure times, as well as through the estimation and numerical averaging of their productivity. In these circuits, agriculture becomes an averaging activity that seeks the control of production, guaranteeing the needs of the markets, as well as preventing the generation of surplus. Embedded inside visual circuits and broadcast over irrigation zones conceived as signal territories, the streams of water, chemicals, and human workforce converge in these agricultural programs as a systematic practice of slicing surfaces of commodified light. Agriculture becomes a circuit of light: it gives rise to assets as it is seen from the air and projected back to the ground. Here the multitude of techniques, from signaling and broadcasting to the different territorial markers and textual management of time, became enmeshed in how surfaces can be programmed. In many ways, this also relates to the long history of the financialization and assetization of agriculture.[75]

IMAGING AND PRINTING THE ENVIRONMENTAL SURFACE

Most of our chapter has been about internal colonization through agriculture and what it meant as a particular circuit of aerial imaging, ground practices, and plant growth in the midst of Franco's Spain. Here, the centralized formats of planning and modification of land are read as themes related to extraction and gathering of data through the INC, which fed into what could be called a reprogramming of the landscapes of growth, echoing later discussions in critical landscape studies and architecture.

However, we also wanted to connect this to a particular aesthetic question about images, which is why we sandwich internal colonization between the aesthetic question about images that impact environments, and the surface operation of environments programmed for imaging. We are, after all, interested in an ecological aesthetic that reads images in close connection with different techniques of reading, writing, and rewriting vegetal surfaces. In this vein, let us return to the scene in Tarkovsky's *The Mirror*, where we began this chapter. After the sound of a train fades away, a doctor approaches the woman who is sitting on the fence. As he sits next to her, the wood of the fence breaks, and they fall to the ground. As he gets up, dazed, the doctor asks aloud: "Look at these roots, these bushes. Did you ever wonder about plants feeling, being aware, even perceiving?" The conversation continues for another brief moment until the man makes his way back into the fields in the background of the picture. At that moment, a gust of air appears across the scene, a wave that stirs up and sets in motion the stems, grass, shrubs, and leaves of the trees in its path. The front of the gust advances clearly from the doctor's position toward the fence in the foreground. Aware of the phenomenon, the doctor stops and turns his gaze toward the camera, toward the fence he has left behind. In relation to the wind in the scene and in relation to Tarkovsky's notion of the "inner life of the shots,"[76] the filmmaker comments: "Rhythm in cinema is conveyed by the life of the object visibly recorded in the frame. Just as from the quivering of a reed, you can tell what sort of current, what pressure there is in a river, in the same way we know the movement of time from the flow of the life-process reproduced in the shot."[77]

The quivering reed is a sensorial proxy that is pitched as a data-observing technique for a river, for the weather, and for various surrounding dynamics. Instead of data as numbering, it is materially placed, read from the forces expressed through this singular detail. One would expect nothing less from Tarkovsky. We could leave it there and establish a connection to the minuscule moments that film is able to capture as particular images, particular observations of environmental proxies, with a poetic formulation of how the world already gives clues on how to read it as cinema before cinema, as images before technical images. However, as in the case of the previous shot, this analysis omits the material conditions used to construct the image that is fixed in the film. In this case, the gust of wind that sweeps across the scene and animates the vegetation was created with the help of a helicopter hired to fly over the terrain to generate the effect of moving air. The out-of-field presence of the helicopter does not necessarily undermine the author's cinematic project, which focuses on the interaction between the image and the observer and image-events *as* observers. However, as in the case of the buckwheat—as much as in the quivering reed—this scheme of presence and absence introduces new elements whose relevance may go unnoticed if we pay attention only to Tarkovsky's words. It is much like the train that, photographed by William H. Jackson between the mountains of a magnificent canyon at the end of the nineteenth century, reveals not only the technological background underlying the fin-de-siècle landscape epic in the United States of the time but also the landscape-transforming forces in which both photography and the railway participate.[78] Similarly, the action of the helicopter in the airburst sequence reveals invisible threads operating on what is seen, radiating from a position of control.[79] That is, while Tarkovsky's film, seen through the prism of Uhlin's notion of vegetal filmmaking, takes shape with the vital rhythms recorded on the photosensitive film, the growth it expresses does not take place in an open field. Instead, it takes place in an environment of regulated conditions—the control of light or the staging of air or water, even the production of wind—that is closer to a greenhouse than to a primeval forest. The fundamental "border zone" of labs and landscapes is not restricted to the field of biology but is present, in varying degrees, in questions of imaging, even in cinema as a particular manufacturing of environments.[80]

The helicopter in Tarkovsky's scene highlights a space of practices from the air that has been transforming landscapes since the beginnings of the mechanization of agriculture, both on the ground (tractors) and in the air (airplanes and other vehicles).[81] Such a transformation has been spurred on by planes spraying chemicals, which finds its apogee in the contemporary digitalization of the rural. Precision agriculture, for example, can be understood as a visual (data) practice in as far as it relies on the complex interplay of imaging and sensing, vegetation indices, comparative reading of images, and different data analytics and machine learning for improved capacities to model and predict crop behavior.[82] These insights relate to understanding practices that take place thanks to the presence of an atmosphere created by the continuous production and circulation of images linked to plant growth. While the inner colonization has been this chapter's main historical case study, we do not intend to hint that the origins of contemporary precision agriculture are found in the Francoist dictatorship or the other large-scale agricultural programs mentioned in the chapter. Instead, we have wanted to showcase how a particular aesthetics of sensing, imaging, and programming of lands also takes place outside of the usual cinematic contexts, that is, places where images are being mobilized to accommodate growth to a space of logistics while omitting the excluded middles of vegetal endurance, human labor, and—as we will see in the next chapter—soil.

5

GROUND TRUTHS: ENVIRONMENTS OF IMAGES

Scale features in many keys and registers of this book, responding to philosophical prompts on the topic in recent years: scale as generative of a "tangle of new relations," scale as stabilization of forms of knowledge and experience, and scale as it is employed in a history of media techniques.[1] One of our narratives concerns the different scales of vegetal growth, paying attention to connecting techniques and themes that link photometric experiments in laboratories with mass infrastructures that sense and aggregate surfaces as data in the form of images. It is one way of narrating what the observation of planetary surfaces (both earth and extraterrestrial) has evolved into: remote sensing infrastructures that picture surfaces. Accurate and verifiable knowledge about the planetary surface has come from massive infrastructure, military operations, and different commercial and noncommercial services, from resource management to meteorological data.[2] At the same time, it is evident that the significance of this surface cannot be captured easily by one method or a term, whether it refers to cultural techniques such as (remote) sensing or specific technologies such as orbital satellites. The meshwork of sensing of surfaces—geographical, vegetal, geological, architectural—is multiplied in the actual products of sensing, such as images, spectrometric data, and other instances where grounds take off, yield information—images and data emanate from the ground and become packaged into orderly units for all sorts of uses of governance. In this sense, while much of our book deals with particular kinds of living vegetal surfaces, the operationalization of the uppermost crust of the earth as a surface managed through cultural techniques of images and calculation goes beyond plants and agriculture to encompass a tight link between sensing, managing, and programmability.[3]

This chapter focuses on the technique and figure of the ground truth used in remote sensing and machine learning. Ground truth has become a foundational figure for the calibration of knowledge systems. However, it has also significantly shifted how actual grounds, such as geographical sites, are integrated into those systems as data. Whereas in the previous chapter we discussed visual agriculture—where media techniques took not only the role of partitioning but of managing the needs and the productions of the soil—this chapter tracks how visual knowledge about surfaces of the planet, landscapes and territories, has shifted to synthetic knowledge about the surface of images. We are interested in analyzing how the notion of ground truth no longer pertains only to the surface of the ground as a geological or geographic reference point, even if it has played a central role in calibrating remote sensing (e.g., by satellite) for minerals and materials to spot their spectral signatures. Instead, ground truth is read through the constantly evolving set of relations among environments of images increasingly populated by the complex of devices, infrastructures, and protocols in earth observation systems. This double aspect of image and surface, image and (literal and conceptual) ground, runs throughout the chapters of this book, and it comes to feature also in different binds between images and data. Here, our focus is on the epistemic power of a particular standardized form of perceptual mechanism.

We start from the assumption that *ground truth* has both a technical and symbolic meaning in how it negotiates relations between images and material surfaces (geographical, landscape, territorial), and their role in various institutional arrangements of epistemic techniques.[4] Starting from ground truth as a figure of knowledge in remote sensing, we build an argument about the synthetic landscapes brought about by experiments within current contexts of machine learning, which we will refer to as "fake geographies," a term already proposed in computer science research.[5] Such speculative landscapes are intriguing experiments in the creative use of machine-learning techniques that deal with geographical datasets. Ground truths shift from their locations as they emanate beyond the ground itself. Surfaces are created, they are even fabulated. They can be seen as elements of fictioning, but with full operational value (imagine misinformation campaigns in real time with geographically shifting landmarks, imagined territories that create falsities for remote sensing systems). As reference data, ground truth becomes read through the "ground" of the synthetic AI images; that is, through how datasets are mobilized in machine-learning techniques for visual ends while implying multiple insights concerning the formats of knowledge as well as the labor of data collection and annotation.[6]

Geographical knowledge starts with producing images through which we see, observe, analyze, and identify. What has been established by decades of critical research is that the relationship between geography and images is heavily overdetermined: the visual and epistemic systems giving a sense of landscape formations are embedded in multiple social, colonial, gendered, and other forms of representational biases.[7] Moreover, this complex role of images in geographical knowledge has given rise to various forms of epistemic transfers addressed in media theory: maps are understood as media and maps mediate cities that are materially situated as part of multiple layers of technologies, new and old.[8] Territories have become "fiducial architectures,"[9] designed to be read by computer-vision-aided machines, as architect Liam Young argues. In other words, machine vision is not merely about a particular set of technological capacities of sensing and processing data but of formatting broader environments and architectures. Of course, not all observation or sensing leads to images being produced. Nonetheless, we are witnessing a range of practices that take the objects or targets of sensing as images for operational purposes. The world becomes an image or, in our case, the various living surfaces, vegetal formations, and geographical and geological territories are sensed as if they were a pattern and an image. Such kinds of sensing and images are supportive instruments for understanding territorial formations. Operations produce images that define geographical knowledge entities. These can include the most (seemingly) inconspicuous practices, such as coloring maps or populating them with place and site names. They can include observing how everyday life is filled with various forms of geographical knowledge embedded in digital platforms for navigation and other purposes. Geographic information systems are the mainstay of practices that emerge through the mobilization of data and electronic communication technologies where the physical and the virtual sign are entangled.[10] At the back of earlier digital technologies, the current version involves the work of data and software platforms as infrastructures and as mediating environments. Platforms such as Google/Alphabet services and other corporations have become a central part of the remote sensing-cartography-digital service industries.

There is also a philosophical reference point for our argument. Jean-Luc Nancy's *The Ground of the Image* proposes a similar shift that troubles a rigid distinction between the figure and the ground, the ground, and its representation.[11] For Nancy, the image already contains a ground. Even if Nancy focuses on classical art history and the philosophy of images that stems from Western art, his has become a useful reference point for considering the image itself as containing the material ground imprinted onto it but also what is being cut into existence by the image. Hence, we bring Nancy's aesthetic point onto the surface of contemporary remote sensing. Nancy writes: "The image does not stand before the ground like a net or a screen. We do not sink; rather, the ground rises to us in the image. The double separation of the image, pulling away and cutting out, forms a protection against the ground and an opening onto it. In reality, the ground is not distinct as ground except in the image: without the image, there would only be indistinct adherence. More precisely: in the image, the ground is distinguished by being doubled."[12] Importantly, this doubling is both philosophical and technical. Grounds rise through images; images turn into grounds in their own right. This argument, even if originally meant in a more art historical sense, can essentially also help us to understand a range of contemporary data and computational techniques and shifts as part of the broader framework of how we engage with questions of AI—such as different machine-learning techniques—as part of operational images: images that do not primarily represent but operate in scientific, military, and other technical systems and institutions.[13]

To push Nancy's point further, we are interested in what it means in contexts of synthetic imaging and data practices, including AI. To this end, the chapter is structured around three key points. First, we outline the notion of ground truth as a particular technical aesthetic of surfaces that emerges in comparison and calibration. We map the shift from the truth of the ground to pattern recognition as a significant transformation that also relates to questions of machine vision and machine learning as techniques of flattening for the task of data production and analysis.[14] Second, we show how the recognition of patterns moves to the building of datasets as they contribute to the infrastructures of ground truths. Third, we look at examples of synthetic geographies as experiments that help us to understand the ensemble of images in which the ground becomes synthesized with meaningful aesthetic and epistemological consequences. At stake in our discussion is the claim that ground truth is read from a mass of images (from datasets that are already embedded in techniques of collecting, comparing, calibrating, and modeling). And in the process, ground truth becomes a mass image, in Seán Cubitt's terms, an aggregation of sensed materials in the contemporary logistics of databases and platforms.[15] We end up with a further emanation of the surface ground that becomes a ground in reference to synthetic abstractions, the sort of circulation of images and data that comes to characterize this second-order set of cultural techniques of remote sensing.

GROUND TRUTH AND CALIBRATION

As a term, ground truth features in geographical and environmental sciences and remote sensing. It is a peculiar linguistic term. While references to truth locate the concept in the seemingly immaterial space of epistemological values, the ground part of the concept alludes to a presumed tangible substrate of firm evidence included in its multiple uses across different philosophical discourses. As Caren Kaplan has observed, "'Ground truth' anchors contemporary preconceptions about physical geography to the comforting solid matter of the earth's crust."[16] The idea of witnessing and proximity is closely related to the epistemological trope of ground truth, which thus resonates with Kaplan's note about the implied solidity of truth, like a permanent and stable geological formation. In more institutional terms, the geological comes in through the work done by various institutions, such as the US Geological Survey, instrumental backbones for establishing ground truth data for calibrating satellite-based remote sensing.[17]

Ground truth seems to imply a direct view based on firsthand observation, implying even a literal ground-level view situated on the earth's surface. But as much of critical science and technology studies (STS) literature would be quick to point out, there is no such thing as "direct observation" that would work as a contrast to a mediated one.[18] Instead, it would be more accurate to refer to ground truth's function in relation to visual and invisual forms of knowledge: ground truths are established to calibrate maps, models, and remote sensing technologies. It is a concept that operates by recognizing a distinction between different sources of data, which, by comparison, can be brought to verify certain features of geographic features, material constitution, or whatever else is being sensed and measured. For example, textbooks such as *Remote Sensing and Image Interpretation* use the term "reference data" instead of ground truth:

> collecting measurements or observations about the objects, areas, or phenomena that are being sensed remotely. These data can take on any of a number of different forms and may be derived from a number of sources. For example, the data needed for a particular analysis might be derived from a soil survey map, a water quality laboratory report, or an aerial photograph. They may also stem from a "field check" on the identity, extent, and condition of agricultural crops, land uses, tree species, or water pollution problems. Reference data may also involve field measurements of temperature and other physical and/ or chemical properties of various features.[19]

As pointed out by the authors, ground truth might thus be nothing related to the ground in a literal sense as it might involve other than solids (e.g., water) or be produced from aerial photographs or comparative laboratory studies (e.g., of the spectral characteristics of a mineral). Hence, as some in the 1970s argued, perhaps instead of ground truth, a more apt term would be "surface observations."[20]

As already briefly hinted, in remote sensing, ground truths help verify, interpret, and calibrate sensing. Much of the work of sensing comes to mean cataloging existing spectral signatures to decode instrument readings, thus necessitating multi-institutional collaboration or intelligence operations such as during the Cold War, even in projects like the Earth Resources Technology Satellite, Landsat. Ground truth, as Pamela Mack explains, was part of the operationalization of infrastructural capacities that built on the idea of synchronization in multiple ways:

> Before Landsat data could be used on an operational basis, researchers had to determine basic "ground truth." In other words, data from a specific image had to be correlated with the actual features on the ground. For example, to learn to differentiate wheat fields from corn fields in an image provided by sensors carried on a high-flying aircraft or spacecraft, scientists first identified wheat and corn fields in an image of a known area where actual ground surveys had been conducted. Researchers had to determine not only the spectrum of light of a particular kind of vegetation reflected but also how this spectrum varied with the growth of the plant, the variety being grown, the character of the soil showing between the rows, and regional variations in moisture and soil. Even the orientation of the rows of a field crop relative to the scan lines of the sensor affected the appearance of a plant field in an image taken by a scanner like that on Landsat. Once scientists worked out methods of differentiating various kinds of vegetation, they still had to develop methods to use the resulting information to predict harvests or to support land-use studies.[21]

Besides technical use, calibration represents an interesting cultural practice in its own right. How, for example, can distances be negotiated and become standardized against a set of features that are assumed to stay regular? To calibrate is to assume standards, to synchronize, to relate to existing accepted details that are the "ground" on which further operations can be scaffolded. While such terms trigger the possibility of long philosophical excavations into the fundamentals of knowledge, the terms are interesting as part of the actual operations of knowledge that work by way of graphic notation systems, inscriptions, and mediations—some of them on the earth's surface. Here, fascinating examples can be cited from the history of remote sensing and the Cold War, when access to foreign territories could not be guaranteed,[22] as well as more recently from the archives of NASA, Google, and many other institutions. Large-scale ground formations, from a set of lines to black-and-white grids and other shapes, mark the surface of the earth as scale calibration models and other test features that function to establish an epistemic base for different sensing, vision, and data systems. The readability of the earth's surface, a crucial feature in much earlier aerial photography, as well as, broadly speaking, in aerial photogrammetry, is now also about the production of active capacities for large infrastructural systems such as satellite remote sensing.

Hence, as far as image atlases are concerned, catalogs of remote sensing test sites are interesting collections of "worldwide test sites for the calibration and post-launch characterization of space-based optical imaging sensors"[23] that also concern the formation of ground truths as specified locations (figure 5.1). The Dome-C site in Antarctica is referenced for its flat, snowy surface and its "temporal stability and spatial uniformity." Another spatially homogenous site is the Mali test site in the Sahara Desert, one among the "Pseudo Invariant Calibration Sites (PICS)," and to the north, in Finland, the now defunct Finnish Geodetic Institute's Aerial Test Range in Sjökulla "contains permanent ground control points (GCP), test-bar targets, and two large (15 meters) dark square abutting a 15m white square given to orthogonal edges for use in determining relative edge response as well as smaller three-bar targets for spatial response assessment." Many more similar kinds of sites could be listed, but this gives an idea of the range of earth readings, existing or created, that not only generate the conditions for calibration but also relate to the notion of operational image as large-scale sensing that is very much about the preparation of landscapes to be operationalized. In this light, it is fair to say that ground truths emerge on location; they are local, specific, and situated to offer a grounding for the network of technologies of sense and location. And yet, the location itself is formulated in the very act of its sensing, establishing again a recursive trait onto the site in question.

PHOTOMOSAICS AND THE STITCHING OF GROUND TRUTHS

Although the term ground truth was not used widely until the 1960s,[24] the topographical techniques of ground truthing were deployed shortly after the invention of photography. Ground truthing refers to synchronizing images and maps, as part of epistemic procedures of verification and calibration, and relating them to techniques such as triangulation,

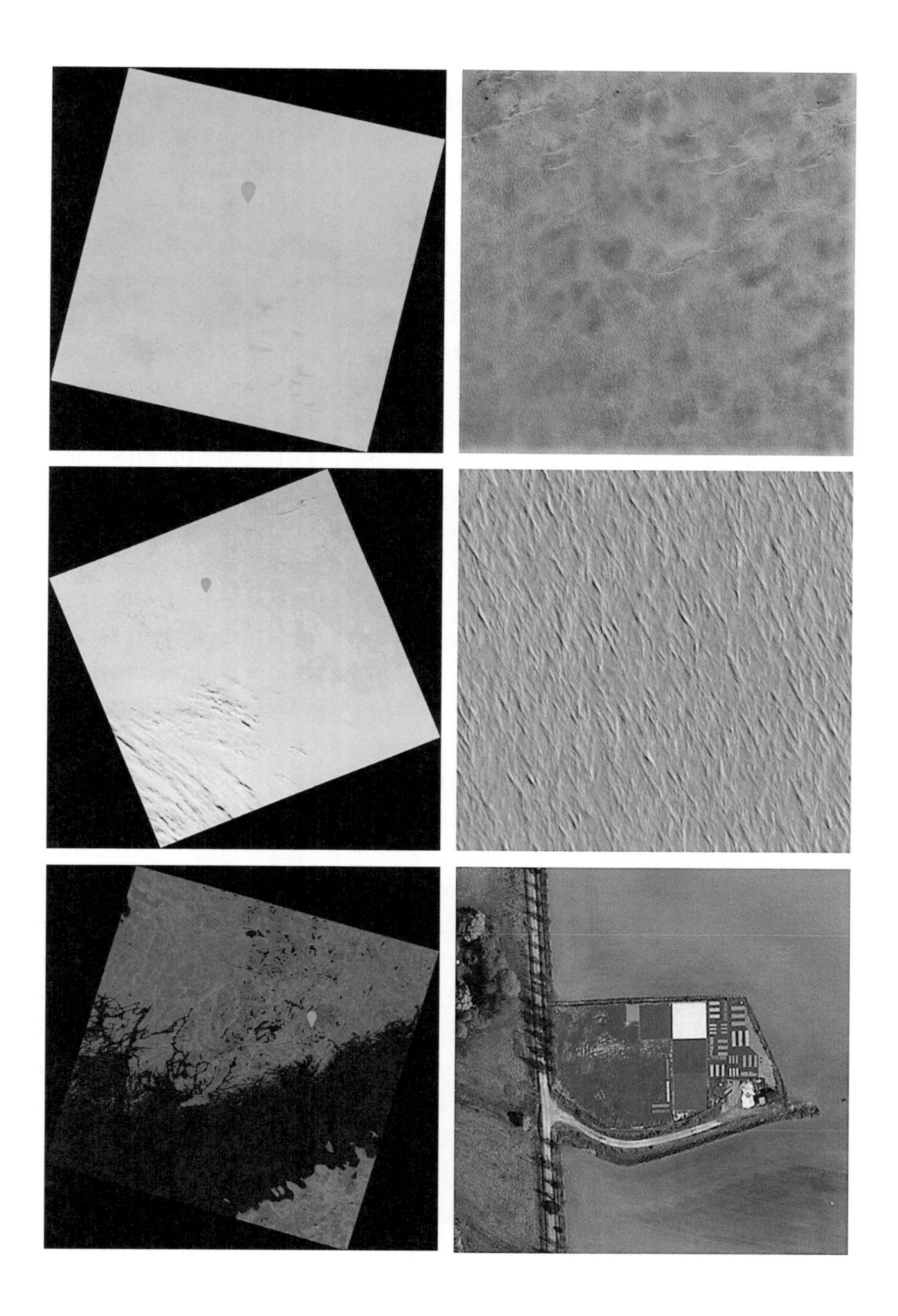

5.1

Landsat 8 and Google Earth Images from the EROS Cal/Val Center of Excellence (ECCOE) Test Sites Catalog. From top to bottom: Mali, Dome C, and FGI Sjökulla Aerial Test Range. Source: https://calval.cr.usgs.gov/apps/eccoe_test_sites_catalog.

as in telemetry. In photography, only ten years after Daguerre's invention, French army officer Aimé Laussedat produced the first aerial surveys, from balloons, which were used by pioneers of photogrammetry.[25] Five years onward, French photographer Nadar filed for a patent on the use of overlapping photos in these surveys.[26] Besides the early photographic and photogrammetry context, we emphasize the centrality of the early twentieth century and the First World War to the operationalization of images of landscapes. As shown in detail by Paul K. Saint-Amour, the military uses of photography after the development of airplanes became central to the early stages of the image-map complex, which radically altered ground truthing (figure 5.2).[27] This historical point about the photomosaic during the First World War will contribute to the main argument we are developing about the synthetic data realms of ground truths, while drawing on Nancy's point that grounds rise up in and through images.

Aerial photography was reliant on a multitude of operations that supported its existence and validity. Ground truth was practiced through this assemblage of skills and techniques, technologies, and their epistemic functions. Besides encouraging the training of personnel to have a fine-tuned capacity to read terrains and images, new technologies supported the task of image comparison and synthetic knowledge. The personnel were a "highly trained interpretive elite," often compared to detectives as they learned to extract as many visual clues as possible from single aerial photographs.[28] In addition, "a complex technological matrix"[29] was set up to help in the execution of this task. This included technologies such as the stereoscope (used with pairs of aerial images) and the hyperstereoscope (an improved version of the first, relying on the constant speed of planes to compare two pictures separated by a known temporal gap) and a specific adaptation of the embodied perceptive skills of the interpreters needed to operate these techniques.[30] With the aid of this reconnaissance matrix, called the "deadliest weapon in the war,"[31] armies were able to distinguish features on the images related to elevation, the third dimension of landscapes, such as differentiating trenches from embankments, as well as being able to see what was hidden underneath bridges and forests.[32] Thus, while the aerial image provided a way of transforming landscapes into readable surfaces, in cases when the dimensions of the

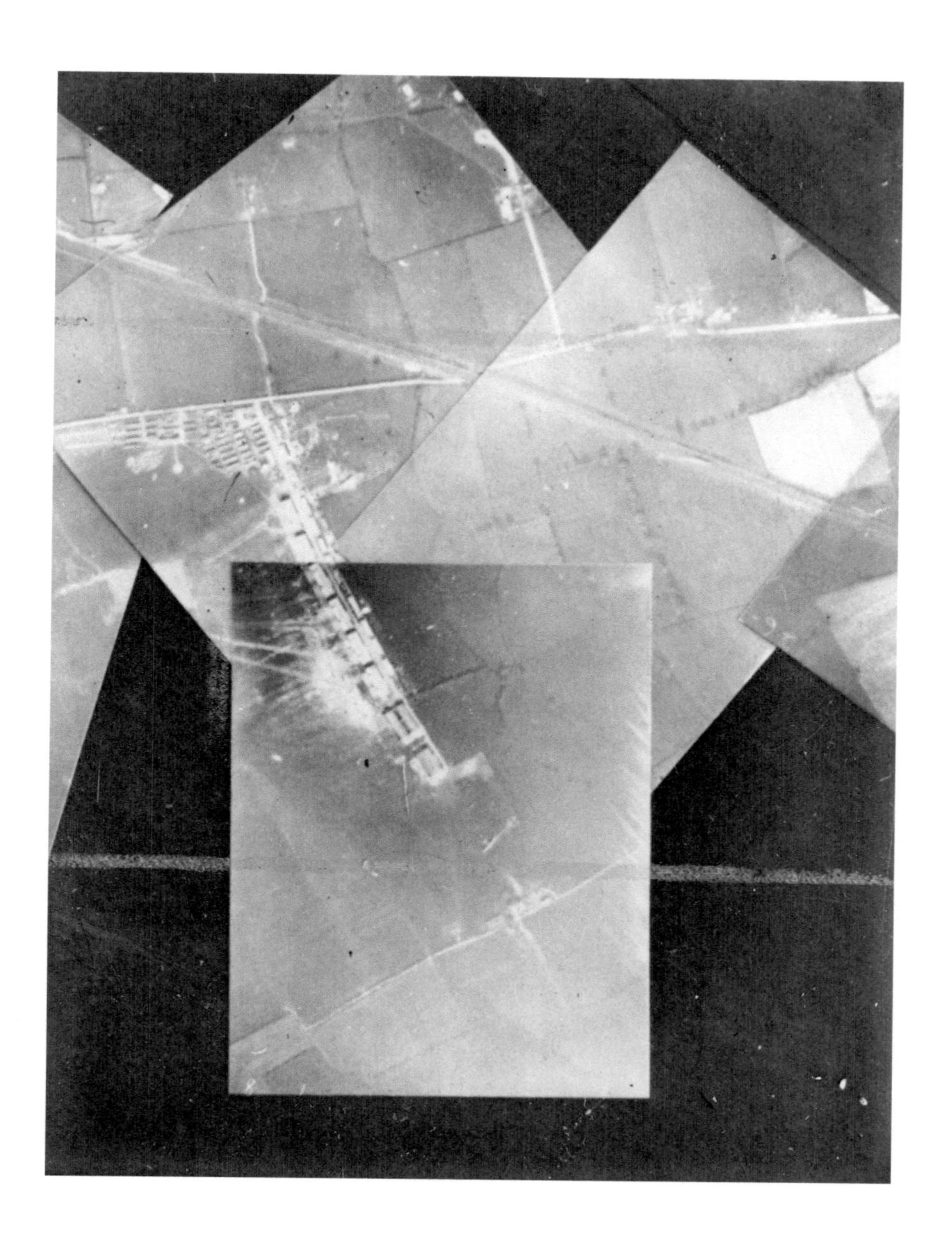

5.2

An example of a composite "photomosaic" photograph of Northolt Aerodrome, London. Source: Imperial War Museums, Q 65971. Image used with permission.

landscape exceeded the representational and encoding capabilities of isolated pictures, simultaneous multiple images were used.

In addition to making elevations visible (in a way, interpreters accessed "a three-dimensional scale model"),[33] other technologies helped to produce pictures of areas on a large scale—such as the areas occupied by trenches—that would have otherwise been impossible to capture in a single shot. In these cases, different images were stitched together to build a large photomosaic. As Saint-Amour showed, technicians relied on the appearance of several recognizable objects in the images.[34] Ground truthing the images, they identified these features as reference data, using them as anchor points when stitching the images together. In other words, techniques of ground truthing had two functions in photomosaics: referential objects were usually tracked to compare and link a particular image to the map of the ground, but here they were also employed to connect images to each other. That is, the ground truthing techniques used to turn images into maps that are useful for navigating and reading the surface of the earth also operated to keep images linked to each other. This is particularly relevant in the scope of this chapter, as it shows how the same techniques were used in two different operations. The same techniques used to verify the correlation between data (images) and ground were also in operation to hold the images together, that is, to keep images connected not only to the surface of the world but also to each other in a meaningful epistemic and operational continuum. It is these operations of comparison, synthesis, synchronization, and calibration that define the scope of ground truth as it emerged as a media technique even before contemporary versions of machine vision and machine learning.

The example of the photomosaic shows how techniques involved in the concept of ground truth were also central to enabling the upscaling of photographic images by stitching them together. Ground truthing becomes—in this case—relational, stabilizing the images in relation to the map, maintaining them as legitimate geographical tools. The epistemic dimension of what is verifiably *out there* is managed through the media techniques mentioned above. Interestingly, this is a relational dimension of the concept of ground truth that has been highlighted elsewhere. Writing about the concept of ground truth as used in the domain of contemporary planetary remote sensing, Jennifer Gabrys argues that "the ground of ground truth is not, however, the final point of resolution in these sensor environments. Instead, it is a reminder of the constant need to draw connections across phenomena. Ground here is connection and concretization."[35] This points to the similar traits we have put forward through the examples above, building a case for grounds based on images and, next, the evidential paradigm that also connects to automation of the interpretation of image-grounds.

WHERE ON EARTH? CLUES, SIGNALS, AND EVIDENCE

It is fair to say that while images are used to stabilize particular ground truths, in the broader context ground truth is itself constantly situated in a set of dynamic processes that fabricate the ground. These processes are technical but are also fundamentally about forms of social and economic power, as some in critical geography pointed out in the 1990s.[36] Already in that earlier phase of geographical research, it was noted that the computerized environments of geographical systems might fundamentally change the nature of ground truth, where the mediations are becoming distanced from the actual ground that was still characterized as a material and lived environment: "The computer promotes a remote, detached view of the world as seen through the filter of the computer database. Intimate knowledge of the world recedes into the background of 'ground truth' as the computer screen becomes the medium through which the geographer interacts with the world."[37] In this context, "the computer screen" was still considered as the central mediating element, even detaching the viewer from reality with a hint of a Baudrillard type of emphasis on simulations departing and distancing from actual reality.

Besides the discussions about the computerization and mediatization of material landscapes and of screens and simulations, ground truth has now become part of the standard vocabulary in machine learning. In this context, ground truth refers to the data provided as sample output values of the training and testing phases in the machine-learning pipeline. That is, it refers to the set of given outcomes—obtained by any means—used to build the model during the so-called learning process. Ground truth does not distinguish between sources of data but refers to a distinction between the outcomes produced by a model and the data provided as comparable expected values. Hence, this incorporation of the ground into the machine-learning operations of producing models becomes a central part of how we understand image operations in AI culture, as an interplay of images, data, and material environments.

Our approach in this book is situated in visual studies and critical (including artistic) studies of data and AI culture, while we also draw on material that has investigated questions of surfaces in both screen culture and environmental approaches to territories. This means that we account for the ways in which material surfaces become not only the object of systems of perception and analysis—like in aerial photography—but how they become its core elements as the ground truth is subsumed in their operative logic, not only analytical observing but the synthetic creation of images from images. This relates to how all sorts of technical sensing are already embedded in the creation of models, calibration, and other techniques that ground what is being sensed. In other words, models are embedded in remote sensing from the first point of contact, so to speak, which makes it at least theoretically difficult to separate a (data) model and data capture through sensors. As Paul Edwards puts it: "Today, no collection of signals or observations—even from satellites, which can 'see' the whole planet—becomes global in time and space without first passing through a series of data models."[38] To name an example: in environmental monitoring, the ground truth of measuring instruments in meteorological stations cannot be separated from the weather forecast modeling practices they are part of. Similarly, seismic data are collected as ground truth as they feed the prediction of seismic movements. Such is part of the recursive nature of many of the techniques and phenomena we investigate throughout the book.

In such domains, ground truth becomes an operation of validating and adjusting a model to a set of facts measured on the ground. While this intervention of the context of the research in the practices of collecting observations has been extensively acknowledged in STS studies, as mentioned above, we emphasize how the notion of ground truth involves an epistemic practice linked to what Carlo Ginzburg named the "evidential paradigm," which emerged toward the end of the nineteenth century.[39]

In other words, ground truth has its own epistemic history as a figure of knowledge.

In the evidential paradigm, the observer—presented under the persona of the detective—is, on the one hand, able to identify a layer of clues and patterns on top of the undifferentiated roughness of matter and, on the other had , to produce a plausible reconstruction of events that have taken place. In this domain, the analytical and the synthetic coalesce. While emerging in a different context than that of visual epistemology or remote sensing, let alone the synthetic technologies of "fake landscapes" in AI—techniques that we will turn to later in this chapter—Ginzburg's insight helps to draw attention to a common trait that characterizes practices linked to ground truth, the idea that ground speaks, or expresses itself, through clues, signs, and evidence that careful observation is able to distinguish, while speculative knowledge emerges from the interpretation and analytical combination, comparison, and aggregation of such traces. Among these techniques are also forensic practices that are part of the contemporary landscape of technical analysis of surfaces and the mediated practices of witnessing, both of which seem to carry forward the evidential paradigm in more (technical) media-specific ways.[40]

In this sense of the term, ground truth applications are found in disciplines such as archaeology, paleontology, and forensic research, to name just a few. In such knowledge practices emerging from remote sensing, ground truth is evoked in various contexts: settlement evidence is checked against images taken from the air,[41] mass graves are unearthed in relation to the indexical appearance of certain species of plants,[42] and agricultural sites are correlated to the detection of phytoliths in soil probes under the microscope.[43] Here the features of verification and calibration play a central role, as do the individuation of signs out of a previously undifferentiated surface and the formulation of an explanatory hypothesis. In other words, ground truth functions as an operation where sets of material traces are distinguished as registers of information while ordered into speculative models.

In all these contexts, ground truth has an interesting relation to Schuppli's concept of *material witness*. This epistemological tool acknowledges not only the evidential role of matter as an active register of external events but also the intervention of explicit acts of scrutiny in the reading of imprints. Here, the "truth" of these "grounds" relies on the application of a series of techniques of demarcation, filtering, and observation, which refer to a domain of practice and a material culture "that enable such matter to bear witness."[44] The forensic method of reading material culture acknowledges that ground alone tells no message; hence we are dealing with "impure matter" affected by the acts of looking.[45] In such contexts where ground truth is established by comparison and other methods, questions of analysis express another take on the evidential paradigm as it becomes involved in advanced computational techniques, including machine learning—and how it stems from pattern recognition.

Not by chance, since the early 2000s, an educational game by NASA, "Where on Earth . . . ?," has been inviting players to become "geographical detectives."[46] The game consists of quizzes where users are asked to locate the geographical area shown on a satellite image by using their abilities to extract visual clues from the image in order to recognize the place (figure 5.3). The archive of quizzes displays islands, mountain ranges, deltas, volcanos, and other geographical landmarks, pictured from above and presented without any revealing textual key. This remote sensing version of a Sherlock Holmes type of analysis would be only anecdotal if it were not for a recent machine-learning project that aims to do something similar on Google's platform, the PlaNet neuronal network.[47] The project aims to build a machine-learning model with the ability to determine the location of a photograph by looking at its pixels, that is, without accessing any image metadata such as GPS information. The image and the geographical area that is depicted are assumed to contain in themselves the keys to their own unfolding; that "context" is already present in the "text," so to speak.

When comparing NASA's educational game to Google's PlaNet, the task's similarity also highlights the main difference: the detective player has been replaced by a process driven by algorithmic big data. Beyond the characteristic automatization of pattern recognition in machine-learning systems,[48] we want to address an additional noticeable difference between the two examples. What is interesting here is an assumption underlying the context of the PlaNet project: any image is supposed to contain enough visual clues and patterns for a sufficiently trained AI model to

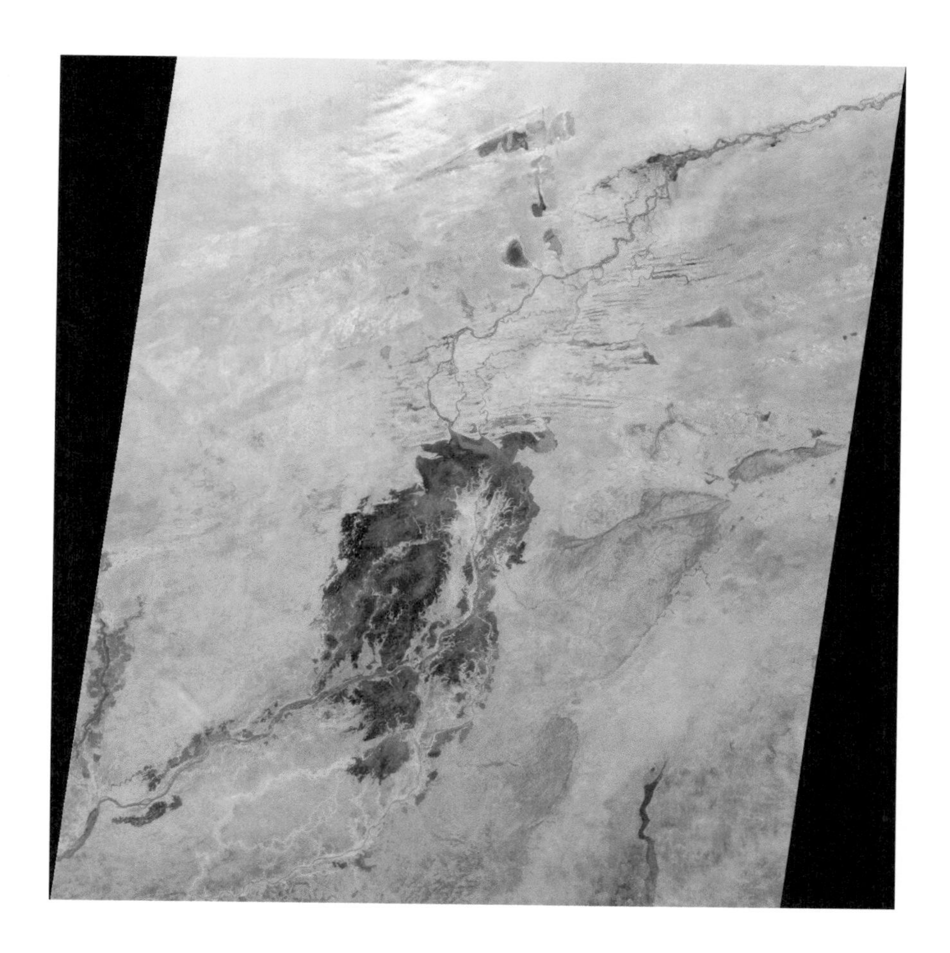

5.3

"Where on Earth . . . ?" MISR Mystery Image Quiz #4. The original quote reads: "Here's another chance to play geographical detective! This Multi-angle Imaging SpectroRadiometer (MISR) image covers an area measuring about 400 kilometers × 450 kilometers and was captured by the instrument's vertical-viewing (nadir) camera on October 29, 2000." The answer to the quiz was the inland delta of the Niger river in central Mali, near Timbuktu. Source: https://photojournal.jpl.nasa.gov/catalog/PIA03429. Image credit: NASA/GSFC/LaRC/JPL, MISR Team.

be able to recognize the place on earth where it was taken. While for the player in NASA's game, the knowledge of geography linked to her pattern recognition skills should be enough to complete the task, in the machine-learning project, the computational model is supposed to be able to identify the place shown in a photograph—once it has been trained with a large enough dataset of all sorts of outdoor images labeled with their geolocation. The physical immutability of the ground in geographical knowledge is replaced by a machine-readable statistical correlation to be found among images when comparing one to another. This is what Adrian Mackenzie and Anna Munster name the "invisuality" of "platform seeing": "Collections of images operate within and help form a field of distributed invisuality in which relations between images count more than any indexicality or iconicity of an image."[49] That is, the ground for geographical detectives is replaced by an invisual ground of relations in the dataset and the context of statistical learning.

The evidential paradigm of the late nineteenth century, observed by Ginzburg, operates algorithmically in the calculations performed among images that build up the AI models and their synthetic knowledge of surfaces. One could go even as far as to claim that the replacement of the figure of the detective coincides with the replacement of the figure of the ground. These different contexts of the use of ground truth unveil a significant transition in the role played by images as part of the monitoring complex of remote sensing of the earth. If in Earth Observation and other instrument and discursive practices, the surface becomes a site of grounding particular truths, these truths highlight how the image complex has evolved and displaced the earth as an object of knowledge to earth data and datasets that constitute the primary reference point.[50]

Next, we move to the question of media techniques of ground truth by discussing another influential project of the past decade that leads us to consider how aerial images are integrated into complex data-synthesizing environments that fluctuate between the visual and invisual. Following Mackenzie and Munster, we argue that this relates to how data are being prepared to be platform-ready and that visual data operate in invisual ways. This relates to the processes of synchronization of visual data prepared for navigational and other purposes.

Continuing the recycling of existing terms in datafied platform contexts, Ground Truth was also the name of one of Google's core projects of the 2010s. It was first publicly described in *The Atlantic*,[51] and the strategic relevance of the availability of detailed and accurate GIS systems in the context of the emergence of all-encompassing digital platforms fueled development initiatives. These Google projects aimed to extract data from images at a massive scale, such as the reCAPTCHA project.[52] Project Ground Truth focused on creating "accurate and comprehensive map data, by conflating multiple inputs, via algorithms and elbow grease."[53] Alongside aerial images, creating access to other combinable inputs—not least Street View cars—was instrumental in offering the synthesis operating at the back of the map services.

The Street View images featured three particularly valuable characteristics: they were regularly updated, accurately geolocalized, and they displayed map-related information such as traffic signs, street names, and brands' logotypes, among others. The Ground Truth Project has been responsible for the developments geared at reading—as in Optical Character Recognition (OCR)—the information printed in the physical world, pictured afterward in Street View images. Notably, these developments involved much more than software engineering, functioning on the coordinated extraction of vast amounts of hours of human cognitive labor—typical of contemporary AI projects.[54] However, the relevance of the Ground Truth Project in relation to the other examples in this chapter is not who or what is in charge of recognizing patterns but, instead, *where* this activity is performed. Significant for our argument about the machine-learning version of ground truth is that the surface of images is the key holder of the information. The key aspect of this case is precisely the circulation, where data from images in datasets are transferred to the images used as geographical maps. The evidence—the clues and signs—are extracted from the surfaces of the images and projected onto the tiled images that make up the map service.

Furthermore, this circulation presents an updated version of how the stitching of images mentioned earlier persists as a key trait of image analysis and machine vision from aerial images to a multitude of data points (such as those obtained from LiDAR imaging), where the ground becomes but a shifting set of techniques that hold together any meaningful relations between grounds and images. Or in other words, it is actually the techniques that fabricate and calibrate the ground. In this regard, drawing on Nancy and other sources, Ryan Bishop argues that "aerial visual technologies and aesthetics . . . are almost solely grounded, literally, in the terrestrial."[55] The inverse is also true when we consider the role of media that establishes the ground as an epistemologically existing composite: the terrestrial is grounded in the aerial (technologies) and, broadly speaking, in the circulation of images.

This is where it is important to remember that aerial photography persisted as a key reference point and infrastructural anchor for the development of remote sensing techniques. Early research on remote sensing in the 1960s showed how the need for geolocating the readings of sensors carried by surveying aircraft—such as spectrometers or radiometers—involved the same hardware that is used in aerial photography. Before the availability of satellite-based positioning systems such as GPS, aerial images were the means to geolocate readings of in-flight sensors, thus connecting "the recorded sensor signals to the ground truth visible in or derived from aerial photographs taken in the course of the flight test."[56] Aerial photographs were shot simultaneously with the sensor's measurements, printing, in some cases, the image of the ground next to the image of the sensor on the same plate.[57] Either through the bare ocular inspection of an investigator, a computer-aided "photointerpreter" with a light-pen, or the operations of an automated system, the sensor data were printed on top of a map or an aerial image of the surveyed zone as if it were a layer of geolocated data in a geographical information system (figure 5.4).[58] Furthermore, the first aerial photograph surveys pictured the ground as well as the sphere and needle of a measuring instrument; each shot of the ground included the image of a compass and clock placed under the camera (figure 5.5). Framed initially by the compass and the clock, the aerial image acquired the role of a navigational tool, though it was later replaced by platforms where images are integrated into broader invisual forms of sensing.[59]

Whether in the current apps of corporate mapping services or in the early examples of aerial images with a superimposed layer of data, the

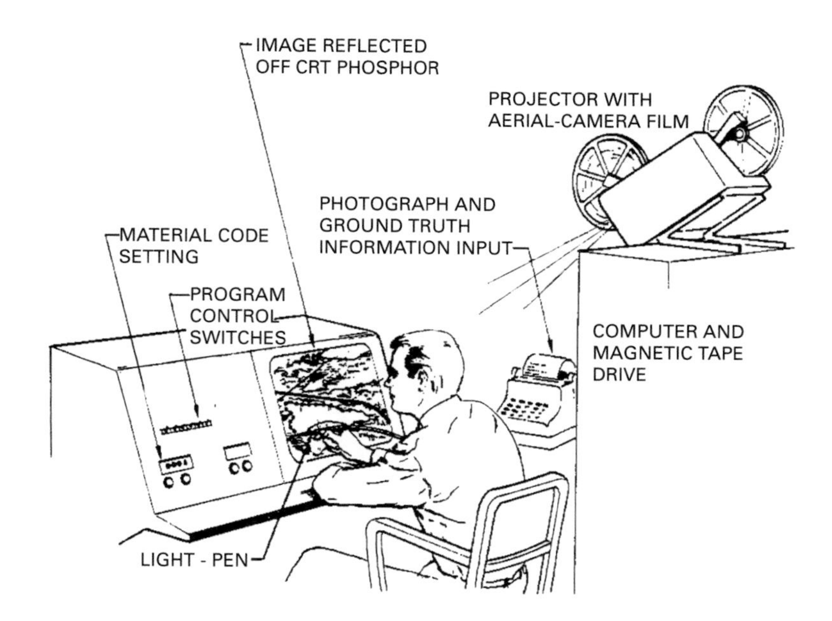

resulting surfaces display data as seamlessly emerging from the ground. However, the process involves a substantial amount of (human) work. This point is crucial in the evaluation of many of the ground-truth projects; the labor of annotation and interpretation plays a central role in the politics of sensing, mapping, and categorizing, as Cindy Lin points out. This also pertains to the broader context of data created from remote sensing (e.g., LiDAR) and how they become annotated, managed, and mediated by data scientists. As Lin argues in her case study of Indonesia's

5.4

Diagram of the editing station with a photointerpreter annotating data observed on the surface of a projected aerial image with a light pen on a cathode ray tube. Source: Walter G. Eppler and Merrill, "Relating Remote Sensor Signals to Ground-Truth Information," *Proceedings of the IEEE* 57, no. 4 (1969): 671. Used with permission.

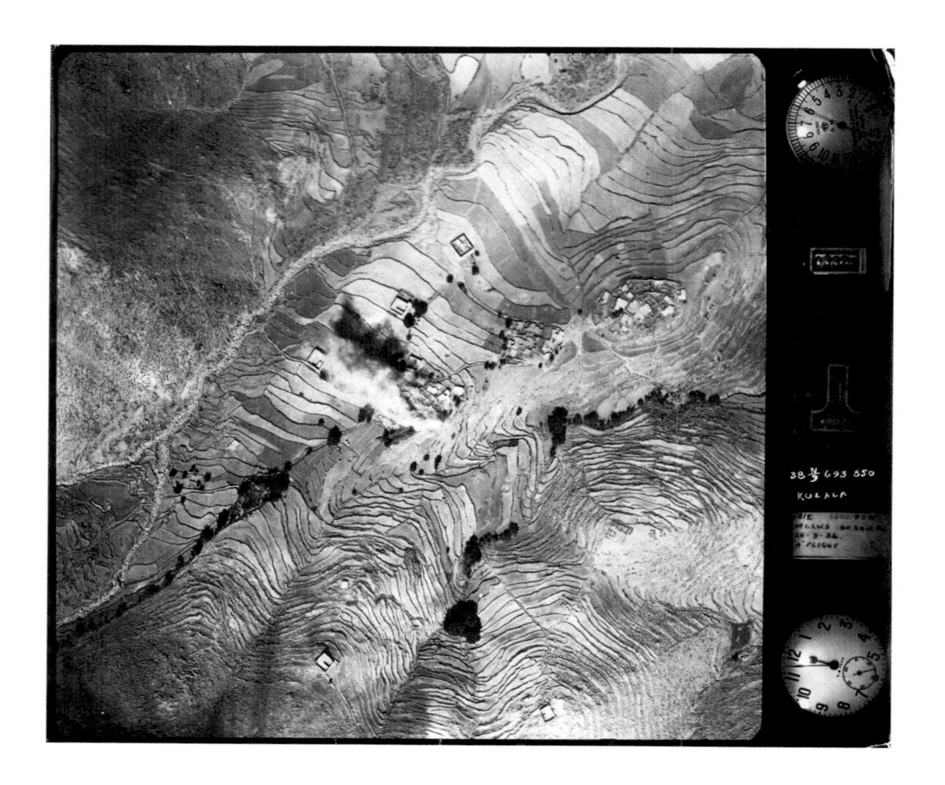

5.5

An aerial photograph as a measurement image synchronized with a clock and a compass. The image is a vertical aerial photograph taken by a Type F.8 aerial camera, showing bombs exploding on the village of Kulala during a raid by "A" Flight, No. 60 Squadron RAF. © Imperial War Museums (HU 91196). Used with permission.

National Mapping Agency, the politics of ground truthing relate to the complex meshwork of how seeing, sensing, and mapping a forest, for example—sensitive due to the question of palm oil plantations—are related to the work of human interpretation of remote sensing images as part of an infrastructure of data from LiDAR to pixel images that become the reference point for analysis. Here, as Lin points out, the issue is about how labor is evaluated in institutional contexts with high political stakes: "Senior bureaucrats characterize data technicians labor as necessary, but likewise inefficient, a mere act of data collection, rather than trained expertise. Bureaucrats have advanced data science projects to automate the delineation of forest borders, to know the forest in precise ways that crowd out what for others is actually there."[60] The grounding images, interpreted pixels, and pixel borders become part of the operational definition of ground truth as reference data are constantly fabricated in loops of referencing. These loops also problematize what is considered remote and what is near, as the political stakes of how the synchronization of different modes of sensing and interpretation takes place are more fundamental.

SYNTHETIC AND FAKE GEOGRAPHIES

So far, this chapter has taken us into the circulating sphere of images that replace grounds. This point stems from a seeming paradox: the use of aerial photographs as a geolocating tool was claimed to project "an image of a de-materializing world,"[61] where spatiotemporal coordinates pinpointing physical ground sites (or targets) were replaced with the circulation of images. While the reference to dematerialization was a characteristic part of the discourse concerning digital technologies from the 1980s to the 2000s, we insist on the material premises of the media techniques at play. In other words, instead of dematerialization there are different aspects of techniques and labor at play that support such image complexes. In this vein, we have analyzed how the concept of ground truth entails a shift from observations practiced at the ground level to operations at the surface of the image. From aerial images to Google Street Views, the priority of datasets becomes emphasized as a core feature of ground truthing that is closely tied to the images' environments. As such, the primacy of images ties together two recent contexts of imaging and ground truth in surprising ways. It reveals something fundamental about such operational images as well as how they are mobilized in contemporary experiments that further shift the notion of the ground truth to fictitious, and even extraterrestrial, land surfaces.

In extraterrestrial remote sensing, we are faced with image analysis where the lack of access to the ground means a complete absence of "absolute ground truth."[62] The exploration of Venus's surface through the images taken by the Magellan mission during the 1990s are peculiarly similar to examples such as Google's PlaNet. A dataset of human-labeled images of planet surfaces—samples of images of craters and other patterns of landscape filtered by expert observers—is separated from the ensemble of images obtained from the mission and distinguished as ground truth for a statistical training and learning process aimed at classifying, at a massive scale, the geographic features on the surface of the planet.[63] In outer space, the ground of extraterrestrial planets emerges from the techniques embedded in the technological infrastructure of orbiting vehicles. Ground truth relies on spacecraft systems, just as travelers and colonizers on their ships carried earlier tools and other instruments.[64]

A case in point is the Mars Exploration Rover calibration targets. These are defined as objects carried in the spacecraft "to verify and validate the preflight calibration" of the imaging systems and "to monitor the stability of the calibration during the mission."[65] For example, the color and brightness of images received from Mars are adjusted in relation to kind of sundial, the Pancam calibration target. Its design recalls the relative positions of the earth, Mars, and the sun (figure 5.6), a fact that is emphasized by an inscription: "Two Worlds One Sun." The sun, in the planetarium-like layout of the components of the target, is represented by the sundial's center. Importantly, it is a sphere that allows scientists to properly adjust the brightness of images of the Martian landscape obtained by the rover. We subtly see in this small detail the echo of a history of moving images when we recall Abel Gance's statement: the future of the movies is a sun in each image.[66] Nancy's doubling of the ground in the image is scaled up to sun as an interplanetary reference point, symbolic and material. Four colored strips around the sundial contribute to keeping

the (images of the) planets interconnected, as they allow the calibration of the pictured colors. This carry-with-you ground truth is emblematic of the point made in various contexts that in such missions ground truth is also portable reference data that has been taken off the earth ground.[67]

Examples like this can be seen as cultural techniques of the complex and costly procedures of extraterrestrial imaging projects and "comparative planetology."[68] In addition to ground truths off the earth ground, we are also interested in a slight twist to the story, provided by different projects that mobilize AI methods to produce synthetic "fake" landscapes. As Bo Zhao and his coauthors outline, while there have been signs of different techniques of AI-driven tactics such as "fabricated GPS signals, fake locational information on social media, simulated trajectories of online game bots, and fake photos of geographical environments,"[69] geographic and territorial (representation) manipulation has emerged as a recent topic of discussion in AI research. Military intelligence's interest in creating manipulated digital representations and mobilizing generative

5.6

Pancam calibration target (left) used to calibrate the colors of an image taken on Mars (right). Source: NASA, Mars Exploration Rovers; https://mars.nasa.gov/mer/mission/instruments/calibration-targets/.

adversarial networks (GANs) into seeing things that are not present or seeing things as misplaced is one indication of the broader applications of fake geographies that invent new relations of ground-image. Such fake geographies recall many earlier versions of paper towns, operationalized Potemkin villages, and other phantom settlements reproduced in different media. However, the automation of synthetic images allows for a particularly different spatiotemporal intensity to the issue at hand.[70]

Deep learning, thus, is mobilized for a range of goals that speak to tactical media at the level of falsified geospatial data.[71] For instance, in such contexts of experimental computer science, the Satellite Image Spoofing project creates fake satellite images, just as deep fakes portray nonexistent faces.[72] Deep fake landscapes shift the focus of both machine vision and AI systems from the individual face as the sole concern of data politics and demonstrate that any image surface—face, landscape, earth, or extraterrestrial—can be treated in similar synthetic ways. It is not only that "grounds" are faked as synthetic creations but that the whole notion and process of synthesis-detection is automated so that reading a representation has to be subjected to automated procedures. It has been suggested that one way of enhancing close reading for detecting fake geographies is to develop computational methods to read minuscule variations of "color, texture and details," as well as "frequency domain features such as a certain type of periodic replications."[73]

In a way, such experimental approaches also resonate with some recent computational art examples. In the *Asunder* (2019) art installation by Tega Brain, Julian Oliver, and Bengt Sjölén, new geographical variations are created and depicted like a dashboard unfolding before the audience. The simulation of imaginary (future) landscapes, driven by machine learning, creates a world for a fictional AI environmental manager to not only observe but reorganize specific locations on earth based on existing environmental data assembled into (at times absurd) projections. While this work is more about the variety of assumptions of rationality built into climate models and projections, it also works with the "machine vision" of fabricating images of terraforming. Techniques of modeling and simulating are part of the same family of speculative design techniques, similar to the peculiar operational frameworks of fake geography.

If we stay with techniques designated "fake geography," the relation between the aerial view and its intrinsic calculability is also explored in generative artworks where images of lands are merged with algorithmic textures, such as in *Neural Landscape Network* (2016) by Gregory Chatonsky or *Invisible Cities* (2016) by Gene Kogan. With similar techniques, Weili Shi's *Terra Mars* mobilized artificial neural networks (conditional GAN in this case) and trained them with topographical data and earth satellite imagery.[74] This model was then applied to see Mars differently, to make it look like earth, as one visual commentary on imaginaries of terraforming and, as per the artist's own words, creative use of AI technologies. Similarly, the Terraformed Mars twitter bot made by the physicist Casey Handmer offers images of "simulated terraformed Mars landscapes every six hours," based on datasets such as that from the Mars Orbital Laser Altimeter (MOLA).[75]

Often, such works are discussed in terms of the creative uses of algorithmic techniques and AI. However, we want to highlight how the image environments are generative, based on data and details extracted and mobilized through comparative techniques. What would it mean to consider these image techniques in this vein? Instead of looking at merely creative AI, we refer to Russian filmmaker Lev Kuleshov's concept of "creative geography" from the 1920s, a concept about building "unique spatial realities . . . out of shots taken in different geographical locations or at different times."[76] This principle of Soviet montage recalls the importance of the relational space opened when ensembles of images are taken together, interweaved in technical operations, and made explicit, for example by Harun Farocki through his practice on the soft montage, something also brought up in our earlier discussion of the *Seed, Image, Ground* (2020) video piece.[77] Hence, the synthetic nature of images and landscapes revolves not merely around current versions of machine vision and the creative use of datasets in deep learning such as GANs but the longer legacy of how techniques of ground truths enable a synthetic creation of truths that, returning to Nancy's phrasing, rise "to us in the image."[78] Indeed, Nancy's take on the ground being doubled and framed in the images becomes an essential guideline—although we must add a further note that it is not only a doubling but a radical synthetic multiplication of grounds that takes place in images as they are mobilized in massive quantities of datasets, the mass-image.[79]

IMAGES AND ENVIRONMENTS ON THE THRESHOLD OF DATA

In this chapter, we have addressed a set of imaging practices related to the production of geographical knowledge. We have focused on an analysis of the techniques as they relate to a broader domain of the image as it approaches the threshold of data. The aim has been to address the shift in contemporary AI culture from the surfaces of the world—such as landscapes and territories—to environments of images. This relation is more than representational, beyond what images depict. Instead, another theme at play concerns how the image-ground relation has been conditioned by a range of measuring, calibrating, and synchronizing media techniques over a longer period. This other realm, let us call it nonrepresentational for now, relates to a shift we have tracked through ground truthing, where the concept of ground truth has been shown to leave the surface of the earth in order to be read through the operations and decoding—and synthetic combining—of image surfaces.

By focusing on the relevance of aerial photography and photomosaics, we have shown how the notion of ground truth has an epistemic history, despite not being mentioned in literature before the 1950s. As a figure of knowledge, it can be traced back to those photographic techniques linked to the first aerial surveys. We have contextualized these with what Carlo Ginzburg exposed as an evidential paradigm, which can also be described by Adrian Mackenzie quoting Hannah Arendt in relation to artificial intelligence: "The crux of the problem rests on the 'treatment' or operations that 'reduce terrestrial sensibilities and movements' to symbols."[80] Following approaches that discuss images' role in machine learning, we have emphasized the importance of the invisual and nonrepresentational domain of relations between images as elements of the data ensembles involved. As per some of the examples in this chapter and earlier work, the point concerns the image being treated through data operations as a surface prepped to give insights into analysis and even synthetic creation.[81]

In this regard, ground truth emerges as a set of techniques where these symbols are related to each other as a media operation that, in addition to grounding, current geographical systems can also use to generate what we have called fake or synthetic geographies. Existing critical work in geography has articulated similar claims, such as those of John Pickles who, writing on geographic information systems (GIS) and "Benjamin's law of assembling images," affirmed: "In this sense, as well as legitimizing claims to verisimilitude, digital mapping signals the end of mapping as evidence for anything, or at least the emergence of a representational economy whose illusions—Baudrillard tells us—will be so powerful that it won't be possible to tell what is real and what is not."[82] But there is more at stake than the earlier discussions about real versus unreal, digital versus analog. Different techniques of ground truth also reveal a different set of epistemic and operational assumptions about surfaces—as targets, as resources, camouflage, and so forth. This builds on the earlier context of simulated territories—even simulated realities if we follow the wake of Baudrillard—but specifies it also in terms of some of the techniques at play in more recent contexts of synthetic surfaces.

Beyond geography and GIS, ground truth has its main domain of application in AI techniques and machine vision, a point that comes through also in artistic practices that address such ideas of the ground as a speculative, calculated, hypothesized entity. These AI experiments in computer science and art help to reframe some of the questions of the supposedly solid and stable, truthful grounds, and they extend the trajectory of the visual culture of the Anthropocene, which is parallel to "the transformation of the world into images," with images "constituting a new kind of knowledge."[83] Thus, the discussion of ground truths and environments of images also concerns contemporary AI-based data cultures when it speaks of technologies and the institutions of the verification of data, including ground truths, reference data, and surfaces across images and planets.

6

LIVING SURFACES OF MEASURING NUMBERS

So far in this book, we have addressed different scales of soil and vegetal life in relation to the emergence and stabilization of surfaces of another kind, namely those of images. Images measure how light enters and reacts on the vegetal surface of the earth. Thus, also images participate in the metabolic processes that light enacts. This becomes even more evident the closer we get to the practices of modulation of light in terms of modern architectures of glasshouses and phytotrons, techniques and technologies from synthetic chemistry to data visualization, and other examples that populate this book. From the microscopic chloroplast to the ensemble of living matter, from leaves to agricultural landscapes, the sentient and terraforming characteristics of plants have been modeled from and through the environments of images. Environments where images are enmeshed as tiled maps of satellite images, datasets for machine-learning models, photogrammetric scans, and streams of networked cameras have reshaped the uppermost crust of the living earth and become a new multiscalar reality. This synthetic surface wraps the planet and alters its cosmic relations.

There is, however, a very specific scale that we have not addressed until now, which will be the subject of the last two chapters of this book. It is the scale of the vegetal communities that plants constitute and maintain on their own; it relates thus to the specific groupings of populations of different species of plants that coexist autonomously and spontaneously in the same geographical area or habitat. For example, we are thinking of the assemblages that make up the different grasslands and forests. As we will see in this chapter, these communities can be considered as having specific processes, functions, and characteristics that allow them to define and differentiate several stages of development. That is, the relationships among species and populations that coalesce in these stabilized associations of vegetation are the object of community ecology. Furthermore, it is a scientific practice involving not only plants but images and practices on top of images as well.

In this chapter, we examine the work of one of the pioneering figures of this discipline in the early twentieth century, namely the US botanist Frederic E. Clements, whose research included a theoretical conceptualization of these *plant formations*—his preferred term—as a new type of organic entity. His work proposed a quantitative plant survey methodology specifically designed to examine the formations. His use of images, statistics, and techniques of counting would already make the work interesting in the context of this book, but we also want to examine how his methods relate to the broader extension of quantitative tools outside the lab into the field.[1] Clements's entwining of experimental practice and theorization makes it a notable reference in the case studies we lay out in the book. While Clements was described by such contemporary ecologists and botanists as Alfred G. Tansley as "by far the greatest individual creator of the modern science of vegetation,"[2] it is also outside the area of research of vegetal life that his work resonates. Briefly stated, as a teaser of what is to come, his quantitative and statistical approaches to the spatial characteristics of plant communities are tightly linked to a specific domain of practices that have surfaced since the availability of satellite remote sensing images. Clements's practice is part of a quantification of living environments and surfaces that has given rise to operational techniques where plant population becomes a biomarker in fields such as military intelligence.

Clements's work can be linked with the systematic elaboration of image-based indicators of surface characteristics, such as the *photobotany* practiced by Cold War intelligence programs. We build on the research by Robert Gerard Pietrusko into such techniques of indication and on how ecological knowledge was integrated into remote sensing operations. We argue that the broader characterization of these ecological living surfaces is a version of what John May has defined as the "managerial surface," the "statistical-electrical control space" that nowadays mediates the surfaces of the planet for the contexts of environmental management.[3] Clements's project of numerically analyzing the vegetal ensembles that comprise prairies or forests took place in the pre-electronic era and long before the role of underground mycorrhizal networks performing their hidden "quiet agreements and search for balance"[4] were studied in scientific contexts.[5] A merging of statistics and image-based techniques allowed him to study the surface of these plant formations numerically, resulting in what has been described as a physiognomy of vegetal superorganisms.[6] Besides the discourse of superorganisms, the involvement of media techniques leads to a parametrization of vegetal forms that anticipates contemporary techniques of *photomorphology*, that is, reading photographic traits as one would read plant morphology.[7]

Clements's experimental methods were devised and put into practice in the vast Midwest prairie of Nebraska, as it was in the early twentieth century, with its smooth shifting mix of grasses. In the grasslands Clements established a connection between imaging and modeling surfaces of the earth in much the same terms as it operates in current environmental observation systems. Starting from grass, we will examine the nuances of this relation, taking into account Clements's work and his background at the University of Nebraska, and we will place the relation in the context of parallel discussions relating early photography to computation. Land cover classification and interpretation bend in multiple institutional contexts, demonstrating a key trait of how ecology and environmental knowledge become integrated into different planetary projects, including geopolitical ones. The knowledge practices that were gradually established from the late nineteenth century to the first decades of the twentieth century are reference points in establishing issues that resonate far outside their original domain; the statistical understanding of living surfaces comes to define one insight into the broader discussions about media techniques of ecological control.

THERE ARE NO INDIVIDUAL PLANTS

What is grass, anyway? What type of collective entity do *Gramineae* plants form? How is it that they come together to create a surface as a continuous vegetal entity, where the limits between individuals blur and disappear? The puzzling dissolution of the individual and the collective that characterizes vegetal life features strikingly in meadows and grass plains. "A plant is a colony," in the words of plant scientist Stefano Mancuso and writer Alessandra Viola, showing how plants trouble basic philosophical categories such as individual versus multiplicity and distinct versus anonymous.[8] The striking headless character of plant behavior also features in works of fiction.[9] "There are no individual plants," explains one of the characters in *Vaster than Empires and More Slow*, Ursula Le Guin's short story about an exploration of a planet fully covered by forests and grasslands.[10] A group of scientists realizes that the entire planet's vegetation constitutes an all-encompassing and sentient entity, perhaps somewhat reminiscent of the planet-brain ocean of Stanislaw Lem's *Solaris* (1961) and predating James Lovelock's point that "life exists planet-wide or not at all."[11] Camping on a vast prairie covered with grass-like plants, the scientists' attention shifts from the noticeable lack of animal life to this other form of habitation on the planet's surface. In the words of one of the characters: "Even the prairie grass-forms have those root-connectors, don't they? I know that sentience or intelligence isn't a thing, you can't find it in or analyze it out from the cells of a brain. It's a function of the connected cells. It is, in a sense, the connection: the connectedness."[12]

While Le Guin's story merits much more than this brief mention, it is this inspiring and perplexing sense of connectedness caused by grass that we want to begin with, alongside questions of ground-level visibility versus underground invisibility. If grass had been analyzed in significant cultural histories from the nineteenth and twentieth centuries, in the second half of the twentieth century, grasslands came to be regarded as cybernetic surfaces and were modeled in systems theory.[13] As an example: the Grassland Biome project—a big-science initiative of the US administration in the late 1960s aiming to create a computational model of a grassland ecosystem—imagined cybernetical environmental management with all sorts of flows of energy and matter mapped as data streams.[14] At the back of the emergence of terms such as "ecosystem," the cybernetic research meetings and agenda after the Second World War started to produce different scales of applications relating to ecology. This included theoretical models, such as G. Evelyn Hutchinson's version of the biosphere in the context of the "calculation of information as the control mechanism for the flow of energy in natural systems,"[15] as well as the Odum brothers's work on ecosystems, information, energy, and their applications to agriculture.[16] Grass was thus one of many ecological elements that were subjected to different modeling and testing in ecosystem approaches.[17]

While Gaia theory emerged from comparative planetology, such as research on Mars,[18] science fiction had already imagined outer galactic grassland ecosystems in the early days of cybernetics. *Green Patches* (1950) by Isaac Asimov features a planet that is described as having a form of grass that does not grow only on the surfaces of soil but also in the eyesockets of the creatures populating it, such as grazing animals.[19] The theme of *grass replacing the eyes* is presented as the ultimate regulator of every single metabolic process in the alien biosphere. As in a cybernetic fantasy, grass establishes the planetary equilibrium by monitoring and controlling each exchange of matter and energy among living beings. Grass becomes seeing while seeing becomes cybernetic regulation and control in an alien version of not only Gaia theory but also smart city imaginaries of feedback loops of sustainability.

PLANT FORMATIONS AS SUPERORGANISMS

Before any such science fiction narratives of cybernetic plant-eyes, prairies of grass—such as in the American Midwest—became a platform for quantification, analysis, statistics, and modeling, Frederic E. Clements, a member of the so-called School of Ecologists of Nebraska, undertook a major study of these vast landscapes of plants, aware that this also meant tackling questions of visibility and (root system) invisibility. His approach combined a series of quantitative experimental methods with an ambitious explanatory framework: the plant succession theory. This theoretical framework aimed to introduce, at the same time, a structure underlying these forms of vegetal zoning and a temporal dimension of development stages meant to explain their variability. That is, extensions of grasses such as the ones surfacing the prairies of Nebraska were examined by Clements as the spread of an elemental unit of vegetation, the plant formation.[20] Plant formations were seen as complex organisms able to reproduce themselves, adapt to their surrounding environments, and evolve through a series of fixed stages. The following description by Clements presents concisely the concept and many nuances entailed by it:

> The plant formation is an organic unit. It exhibits activities or changes which result in development, structure, and reproduction. These changes are progressive, periodic, and, to some degree, rhythmic, and there can be no objection to regarding them as functions of vegetation. According to this point of view, the formation is a complex organism, which possesses functions and structure, and passes through a cycle of development similar to that of the plant. This concept may seem strange at first, owing to the fact that the common understanding of function and structure is based upon the individual plant alone. Since the formation, like the plant, is subject to changes caused by the habitat, and since these changes are recorded in its structure, it is evident that the terms, function and structure, are as applicable to the one as to the other. It is merely necessary to bear in mind that the functions of plants and of formations are absolutely different activities, which have no more in common than do the two structures, leaf and zone.[21]

The different types of prairies, forests, deserts, and Arctic landscapes, among other landcover types, are characterized as the habitats of a class of organisms distinct from that of the individual plant. Plant formations grow, evolve, and reproduce themselves, giving rise to landscapes of vegetal zoning. These are occupied by different types of formations involved in functions such as association, invasion, and succession—in other words, functions geared to maintain their own collective metabolism, spread into new areas, and reproduce themselves.[22] In spatial terms, the idea of plant formations allows for the projection of a sense of structure among the vastness of some of those vegetal surfaces. In relation to time, it introduces a vocabulary of types and stages of formation that accounts for the dynamic variability that an expert eye could observe inside those extensions.

Clements's ideas about plant formations are also linked to the terminology of "superorganisms," which is one of the reasons his work is still discussed, including within the context of Gaia theory and Anthropocene debates.[23] Indeed, Clements's work seemed to imply that the whole of the prairie could be, hypothetically, seen as an organism in its own right, although the empirical side of the work was more precisely defined. As Clements writes, "As an organism the formation arises, grows, matures, and dies. Its response to the habitat is shown in processes or functions and structures, which are the record and the result of these functions. Furthermore, each climax formation can reproduce itself, repeating with essential fidelity the stages of its development. The life-history of a formation is a complex but definite process, comparable in its chief features with the life-history of an individual plant."[24]

Clements's view of vegetation as an organism where the sum is more than its parts resonated with similar accounts in animal research. Some years earlier, in an article from 1911, William M. Wheeler had referred to the idea of "the ant colony as an organism,"[25] while in the 1920s, biological and ecological research on holism drew directly from thinkers such as Henri Bergson and Alfred N. Whitehead.[26] In 1926, Jan Smuts published his influential *Holism and Evolution*.[27] In all the accounts across the 1920s and 1930s, with different emphases and reference points, earlier mechanistic notions of life were replaced by descriptions of a vibrant relationality and ecology of interconnectedness as the driving forces of life. Furthermore, while critical of calling plant formations *organisms*, let alone *complex* organisms or even *communities*, Alfred Tansley acknowledged how Clements's solo and collaborative work contributed to ideas of quasi-organisms of plant life, and influentially, even a focus on "systems" that included organic and nonorganic parts in interaction.[28] Tansley was far from embracing the holism, let alone vitalism, of some of the contemporary thinkers of the 1920s and 1930s, and yet the lineage of succession, interaction, and complex relationality remained an important part of the methodology during the subsequent decades, when his notion of "ecosystem" continued to be picked up in the later era of cybernetics and systems theory.[29]

Clements's idea of a vegetal superorganism faced critique in the 1920s, perhaps with the underlying fear that the notion sounded too mysticizing and risked conflating complex organizational entities such as those in human and animal life with the way plant life developed. Yet the notion was not necessarily meant to refer to a horizontal organization of entities living in peaceful symbiosis. As Thomas Kirchhoff argues, the thesis should not be read as comprising "mutualistic unities but as multilevel top-down control-hierarchical unities."[30] Here the idea of a dominant formation as actively shaping the broader ecological context of life becomes a central part of this hierarchical way of understanding organicism: "A dominant is an organism with such definite relations to climate and such significant reactions upon the habitat, or in water upon the other community constituents, as to control the community and assign to the other species subordinate positions of varying rank."[31] In other words, the dominant formations are not merely expressions of the landscape surroundings and ecological relations but they also reform them: "It is an axiom that the life-form of the dominant trees stamps its character upon forest and woodland, that of the shrub upon chaparral and desert, and the grass form on prairie, steppe and tundra."[32] This description does not reduce the interesting aspects of Clements's ideas—that the activity of formations is irreducible to taxonomic categories—there is still at play a dynamic notion of a unified mechanism of plant community, even if this synergy is not entirely horizontal and mutualistic but defined by a rhetoric that sounds like some later realizations about control, feedback, and hierarchy found in Cold War–era cybernetics.[33]

NUMBERS: A MINUTE STUDY OF THE UNSEEN CHANGES

In addition to the discussion of plant formations as living organisms distinct from individual plant specimens, one of the most salient characteristics of Clements's work was his continuous drive to think quantitatively. He relied on existing instruments, such as the automatic ones he recommended, for periodic measurement of the physical factors surrounding a specific stage of any plant formation—such as humidity, light, temperature, and wind.[34] But in addition, he had to design new techniques when no other options were available. This was the case in particular in the measurement of the spatial structure of a formation and its relation with adjacent formations. Clements's view of formations as organisms, together with his insistence on quantitative methods, got him involved in the development of specific techniques of plant surveying precise enough to provide the researcher with "a minute study of the shifting and rearrangement of the individuals."[35]

Until then, quantitative plant survey methods had dealt with the spatial and geographical dissemination of plant species, that is, plant geography. At the back of colonial travel and logistics, access to different regions allowed new comparative perspectives and even quantified analysis to emerge. Beginning in the nineteenth century, for instance, sponsored by the Spanish crown, reaffirming its grip on its American colonies, Alexander von Humboldt pursued the first major step in quantification in plant geography.[36] His experimental practice, described by Marie-Noëlle Bourguet as part of a material culture devised to arrive at a "physics of plants," relied on the estimation of ratios of species and genera in a landscape and their subsequent correlation with climatic indexes.[37] The isotherm line, for example—a visual medium consisting of a line on a map connecting points of equal mean temperature—became a famous comparative graphic instrument that concatenated statistical variations across vast geographical areas into stabilized regions to show potential comparisons. These regions were plant formations, such as forests, grasslands, and deserts, that divided the earth's surface into identifying categories, even if somewhat ideal.[38] Humboldt's system allowed us to visualize planetary patterns of distribution of flora, although it was useless for territories of a smaller scale. It was unable to provide the "minute study" required by Clements of the smaller variations that ecological areas exhibited across seasons and life cycles. The situated details were missing.

Aware of the limitations of Humboldt's system, a subsequent German school developed an alternative framework of phytogeography, which was gradually improved until the end of the century. The main technique was the partition of a space into a grid of squares or rectangles of the same size, a technique related to earlier practices in the context of German forestry.[39] The size of the rectangles was large—roughly 20 kilometers squared—and the methods for the quantitative estimation of the abundance of species were imprecise. Surveyors relied only on "visual impressions"—expert panoramic views of the landscapes—and on a simple scale of five degrees of the abundance of species, which increased from "scarce" to "sparse," "copious," "gregarious," and "social," with "social" being the characterization of grasses in a prairie.[40] At the end of the century Clements and his colleague Roscoe Pound applied this phytogeographic framework to the habitat they wanted to survey, the vast Midwest prairie in Nebraska. When starting to apply it in practice, however, they soon realized the deficiencies of the method.

Visual impressionism, the sensory approach that this methodology advocated, was deceptive: plants and flowers that seemed dominant to the eye were actually not the dominant ones. Even the trained eye would be likely to miss a plethora of significant traits that remained under the threshold of (qualitative) visibility. In addition, an even bigger issue made the method useless for the context of the prairie: estimations by naked eye were unable to capture the slow shifts and transitions from one type of grass to another. Vegetal formations shaded in and out and remained imperceptible to this manner of surveying. This "silent and insensible shift"[41] was part of the core epistemic concern, which necessitated methodological solutions beyond the register of visibility. Needless to say, the unnoticed changes in grass types were here an epistemic and aesthetic dilemma that should raise implicitly or explicitly the question: Unseen or invisible to whom? The settler colonial scientific gaze? To the original inhabitants of the lands? Under which kind of visual regime is the depiction of visible or invisible, chaos or order being inscribed?

In any case, this unsolvable difficulty regarding the transitions and the problem of tracking the shift between different prairie grass formations led the researchers to their own method, which relied on the "actual enumeration" of the individual plants.[42] For the sake of accuracy and more objective survey results, the impressions by eye needed to be replaced by quantifying techniques of counting in order to track the imperceptible shifts in the meadows. Counting individual plants across a sample patch was meant to provide new objectivity, not by means of mechanical instruments only but through techniques of freeze-frame sampling and statistical analysis. As we will see next, this counting was, in principle, the same operation that photography had enabled. Photographic surfaces also functioned as sites of numbering as they enabled second-order analysis to emerge from what was captured in the first place in the image-cum-sample. In Clements's experimental practice—notably, he was the son of a photographer—images and enumeration coalesced on the same surface.

NUMBERS AND PHOTOGRAPHY: A BRIEF INTERLUDE

Before specifying how this enumeration operated for Clements, let us remind ourselves how images, calculation, and computation had already featured in discourses of early photography. Providing one reference point for our argument about the transformation of visual technical media, Geoffrey Batchen has observed how photography early on was already linked to the idea of counting and computation in what he named as the conceptual convergence of photo-media and computing in the 1830s.[43] Batchen builds his argument about early photography involved in computation by reading some of the interactions between Charles Babbage and William Talbot, two pioneers of computing and photography, respectively, and also "expert mathematicians." As Batchen argues, reading Talbot's point about light and shadows, the informational economy put into place was premised on the variation of those two binaries—light and shadow—in ways that established the "system of representation involving the transmutation of luminous information into on/off tonal patterns made visible by light-sensitive chemistry."[44] Focusing on Talbot's 1845 "Lace," plate XX from *Pencil of Nature*, the haptic touch of lace on the contact print of photographic paper represents not only a material aspect of this process but also allows for the image to emerge from the play of presence and absence (of objects). The details of minuscule threads become, in Batchen's parlance, "a ghostly doubling" and an informational patterning of details—even geometrical units—as in the lace.[45] Photographs were seen as tools of counting with a precision that enumerated details in the objects on the surfaces of images. As for their epistemic value, photographs did not "necessarily offer a truth-to-appearance" but "truth-to-presence."[46] Instead of being only visual objects, images were addressed as spatial data arrangements. In Batchen's words, photography involved "an abstraction of visual data, a fledgling form of information culture."[47]

For Talbot, the ability of photographs to "introduce into our pictures a multitude of minute details" characterized the new medium, beyond the lace, as the particular informational unit of depiction.[48] For our purposes of tracking the green surfaces information economy, Talbot's words were suitably accompanied by a photograph of a haystack (figure 6.1) where

6.1

William Henry Fox Talbot, *The Haystack* (estimated 1841). From the collections of the Met Museum, Gift of Jean Horblit, in memory of Harrison D. Horblit, 1994. Public domain. https://www.metmuseum.org/art/collection/search/289179.

the straws are visible, producing one example of this epistemological and aesthetic pairing: how to count through an image?

Instead of the conundrum of individuating pebbles of sand from the grain of a photograph, a more apt assessment in light of the media historical cases we are discussing would be counting blades of grass and hay.[49] It would be impossible, and probably unbearably tedious, for any human to individuate and copy all the blades faithfully in a large-scale prairie of grass. However, photography set the ground for the possibility of doing such tasks by other means. As writer Oliver Wendell Holmes Sr. put it in 1859: "Theoretically, a perfect photograph is absolutely inexhaustible. In a picture you can find nothing which the artist has not seen before you; but in a perfect photograph there will be as many beauties lurking, unobserved, as there are flowers that blush unseen in forests and meadows."[50] In most cases, the epistemic significance of photography went beyond artists or photographers being primary eyewitnesses, let alone counting clerks; instead, it was the surface of the image itself that became defined by such an abundance of information. Photographic images would be "data mined" in multiple ways afterward.[51]

Beyond a characteristic analog-media belief in the possibility of full recording, which Babbage cherished, this fantasy of microscopic detailing of abundant scales of information surrounding photography set the ground for a further set of techniques that grew out of these discourses and material practices. In other words, counting, measuring, finding patterns, and statistical inferring leaped into the domain of the unseen flowers in the meadows mentioned by Holmes, and they also leap into the contexts of prairies made visible and countable by means of Clements's decision to use predefined quadrat areas as units of geometric sampling.

THE QUADRAT METHOD: STATISTICS AS VISUAL TECHNIQUE

This brief interlude—to remind us about the media archaeological link connecting photographs, enumeration, and a representational love for details—invites us to revisit what was taking place in prairie ecology. In particular, it looks at the tension embodied by surfaces of grass between the indistinguishability of individuals against the striking homogeneity of the compound body and the need to perform "a minute study of the shifting and rearrangement of the individuals."[52] This tension is resolved in Clements's and Pound's work thanks to a specifically designed methodology where images and statistics coincide.

While such attention to detail would seem to echo Lorraine Daston and Peter Galison's work on the photography-based notion of mechanical objectivity,[53] counting individual plants in this context emerged from a different background: the use of statistics in the Nebraska School. Different from the work on visual statistics as it featured, for instance, in the charts designed by W. E. B. Dubois and his students for the Paris Exhibition in 1900, Clements's and Roscoe's approach to a quantitative physiognomy of plant formations was related instead to the way "statistics translated into images," in other domains such as criminal photography.[54] As part of their education in the University of Nebraska context, Clements and Roscoe were introduced to the applications of statistics in the context of plant geography. They had also assumed as one guiding premise of Darwin's key idea that populations of individuals evolved rather than individuals.[55] The shifts in the grass were to be read as shifts in population, and counting was the core technique when dealing with population figures.[56]

The replacement of the expert eye did not involve a replacement of the surveyor. Quite the contrary, surveyors were needed to solve how to count individual plants of grass on the visually limitless prairies of the Midwest. This is the same problem we addressed earlier in relation to photography's "minute attention to details," but now surfaces of vegetation were approached as though they were an image of sorts: framed and quantified. In this regard, the solution was not photographic in an analog sense but involved a completely different approach to visual technical media that became prevalent much later. The statistical approach relied on both concrete sampling and mathematical powers of abstraction.

The researchers knew that instead of counting all the individual plants in a valley, they could work on a selection of samples of terrain and infer from them the bigger picture. This way, taking statistics on the one hand and the phytogeographic grid on the other, they arrived at the method that is still used today when teaching about plant surveys—the quadrat method. The method consists of setting up stations around which several squared frames delimit sample areas (figure 6.2). Described in Tobey's words:

6.2

A photograph of a quadrat, showing the process of sampling as image framing as well as gathering data. From Frederic E. Clements, *Research Methods in Ecology* (Lincoln, NE: University Publishing Company, 1905), 168–169.

All the plants within the square were counted and their locations plotted on a grid-graph representing the square. In practice, the scientist used meter tapes along the sides for accurate location. A fifth meter tape was run across one side. Individual plants along the fifth meter tape were counted and located by Cartesian coordinates. When all the plants along the tape had been located, the tape was next moved down the quadrat a few centimeters for counting and plotting the next line of plants along it. This procedure continued until the entire quadrat was counted.[57]

The quadrat describes a method for manually framing and scanning a sample of land, usually 4 square meters. The method involves setting up a station within which a set of physical and meteorological variables are to be measured, as well a number of randomly scattered quadrats around it that would be manually turned into spatial arrangements of information. By doing this, in Clements's words: "The deficiencies resulting from the small size of the plots are corrected by taking a large number of plots at each station and averaging the results."[58] Importantly, it seems that Clements and Pound were actually surprised that the results produced by this method "turned out to be very different from naked-eye estimates."[59]

With the aid of this statistical sampling of the terrain, the researchers could keep track of the otherwise invisible transitions of grass types and calculate averaged plant population data around each station. The continuum of the prairie was this way parsed as a spatial array of data stations measuring plant population shifts and ratios as well as environmental parameters.[60] Photography was in the mix of the instruments that were employed to observe and to measure, as images of the quadrats also needed to be taken for later analysis and verifiability. "No ecologist is equipped for systematic field investigation until he is provided with a good camera,"[61] Clements emphatically pronounced. In his book on method, he offered details on the types of cameras, lenses, tripods, exposure times, and developing techniques required to obtain adequate pictures—even zenithal ones—of the quadrats.[62] Exhaustive and clear, these could be read as a quick introductory course to field photography.

Clements also described what the photographs should look like, adding a further data layer about which kind of media made the living surface addressable: "The ecological view should be a picture as well as a map, . . .

when one must be sacrificed, artistic effect must yield to clearness and accuracy, i.e., technically speaking, contrast must give way to detail."[63] Again, the meticulous attention to detail comes to the fore; this time, however, it is layered across the data image surface (figure 6.3), which became a primary unit for statistical analysis. This offers an insight into a particular technique of both description and averaging that brings enumeration early on into a connection with the framed territory and the photographic image. In other words, these were images of populations on top of a data layer made up of populations—of plants, of quadrats, and of photographs of those quadrats—offering a view to appreciate the image on top of the image: the territory as a site of growth, images as registering these frames of populations of plant life, and the mathematical methods as building these images into useful models.

NEAR SENSING

The curious case of quadrat methods in early prairie ecology shows how shifts in vegetal surfaces turn from invisible to visible both as images and as enumerated samples. The two merge. On the one hand, vegetal areas capture photosynthesis in different intensities and become numbered images—whether through digital techniques or even, in predigital, premachine vision versions, by identifying and counting prairie plants through sampling. On the other hand, different "pixels"—whether on the surface of the image, as part of landcover indexes, or as quadrats that sample 4 square meter pieces of land—are determined not as visual representations but as quantified occurrences and values, where identifying edges and the change of type become the crucial epistemic concerns.[64]

Vegetal surfaces have been interpreted using comparative perspectives of inference and proxies, harnessing the idea of plants as sensors and indicators. Enumerated, computable, indexed, and searchable surfaces are examples of how contemporary management of the earth takes place. The earlier shift to cybernetic management of ecosystems is perhaps one phase in this larger scheme of techniques for making environmental changes visible. The current planetary computational technologies—as well as imaginaries—of digital twins, AI models, and

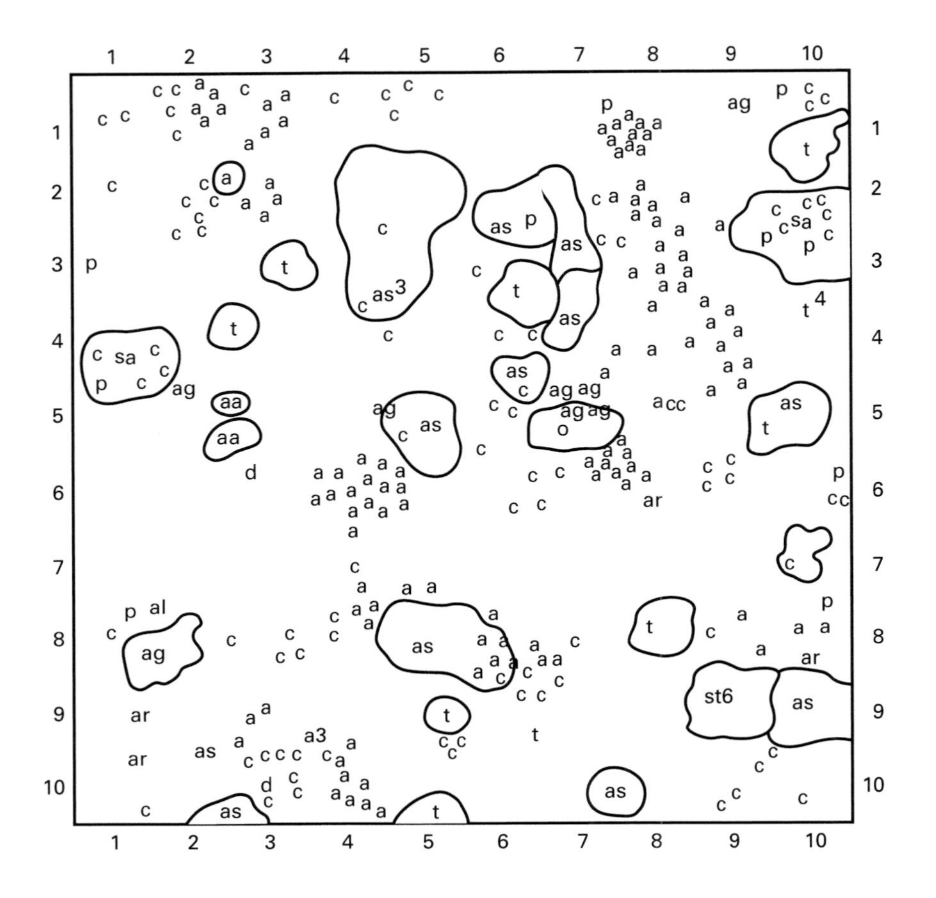

6.3

A "map" view of a quadrat showing the process of sampling as image framing as well as gathering data. From Clements, *Research Methods in Ecology*, 168–169.

prediction environments are echoed in this instance by a quote by Microsoft's chief environmental officer, Lucas Joppa: "Imagine if we had a planetary computer that could tell us exactly what we needed to do to protect planet earth—a system that was capable of providing us with information about every tree, every species, all of our natural resources."[65] As Shannon Mattern points out, the corporate-epistemological dreams of countability, algorithms, and dashboard governance drive many of the current versions of environmental management. Here she is echoing Jennifer Gabrys's point about such plant data as "recourse to metrics that in turn legitimate specific technological interventions to meet targets for averting environmental catastrophe."[66]

Similar elements also define the remote sensing imaginary that takes place across different institutional contexts, from military to scientific earth observation, including sustainability programs.[67] The trope of the surface of the earth as a searchable and manageable database includes governmental remote sensing and earth observation agencies as well as private environmental intelligence companies providing services to use this imagined database. Not that all of it was merely imagined; one could do a whole study of how it is actually the different databases that lie at the back of projects rather than photographic images of the "whole earth" that aim to instantiate the addressability of the earth as a target, as resources, as the digital environment.[68] But such databases and software environments are always much more limited in scope and ambition than the broader discursive production of imaginaries. In this regard, as shown in the next section, when geospatial intelligence companies such as Descartes Labs proceed to build a "living, breathing atlas for the world,"[69] they are not only piggybacking on the historical reference to atlases as ways of organizing knowledge about nature into instrumental collections, they are also finding ways of dealing with surfaces as continuous exchanges between matter and information, the surfaces being both physical and chemical, as well as images, such as those produced through remote sensing, datasets, and data visualization.

Counting vegetal surfaces does not emerge only in remote sensing but in "near sensing," too. Our argument about grass—and counting grass—may be peculiar in the context of statistical techniques of defining and managing surfaces, but it also aligns with the longer genealogy of the green mantle that we referred to above.[70] This quirky choice of an entry point to questions of surfaces, collectivity, and enumeration becomes clearer as we address this particular characteristic that so prominently features in the grass and vegetal surfaces as they are viewed, surveyed, analyzed, and articulated as objects of knowledge and as living, dynamic, interconnected systems. The continuity of grass grounded and stabilized a series of techniques and conceptualizations closely linked to current remote sensing practices in environmental sciences and landscape design: the interweaving of statistics and (what are traditionally considered as) images. While in the previous chapter, we saw how the techniques of building environments of images entailed a transfer from the ground to the surfaces of the image, now we address how continuous slow changes on the large-scale vegetal surface also traveled from the prairies to quantitative descriptions that resulted from networked monitoring stations.

While moving toward this contemporary context where AI techniques and statistical methods are central to counting vegetal surfaces in both urban and nonurban areas,[71] we address grass as a milieu that has given rise to an assemblage of techniques from data sampling to statistical inference as an early example of a managerial surface, which starts to include the key characteristics of computation such as addressability.[72] As briefly outlined in the introduction, while May focuses on the electronic space, the concept of the managerial surface has interesting possibilities in dealing with such images that are not perspectival or representational so much as statistical. What for May is the Cold War–era birth of a "fully automated electronic surface"[73]—defined as "a statistical representation of the continuous surface of the ground, by a large number of selected points with known *xyz* coordinates"—becomes in our focus also part of the predigital, pre-electronic age of operational images that start to govern how we understand environmental phenomena, such as the distribution of ecological life of plants across surfaces of soil, as we have seen in this chapter.

PLANTS AS LIVING MEASUREMENTS

There is an interesting twist in Clements's account that has significant ontoepistemological consequences for how we consider sensing. While his work established both the sampling methods and the categories that define the different forms and evolutions of plant formations, Clements argued that these epistemic categories were not only second-order projections. That is, the plants themselves organized their distribution into zones as they expressed the ecological context. In Clements's words, reflecting this commitment to a realist notion of scientific work: "Zonation is the practically universal response of plants to the quantitative distribution of physical factors in nature."[74] This view of "zoning" thus starts from the activity of the plant, not the mind of the observing scientist: the plants themselves are sensing and inhabiting surface and subsurface areas in ways that also give shape to meaningful classifications. Through plant formations, vegetal surfaces become a primary force that divides space, creating particular ecologically meaningful couplings and inviting thus all sorts of references to "superorganisms" or, in the more modest terms of Clements's later critics such as Tansley, "quasi-organisms."[75] Furthermore, these are not just any plain soil grounds or visible surfaces. The landscapes surrounding plants are *numbered* landscapes; they are quantitative distributions of physical factors, not different from the managerial surfaces we have just described as coming out from the near sensing—statistical sampling and counting—of ground.[76]

Plants' primary dividing force is addressed in Clements's work in relation to these surfaces of numbers. In chapter 2, we saw Julius Wiesner describing the growth and shape of plants in relation to the numbered volumes of light intensity resulting from the practice of the photometer: Clements's reasoning follows a similar path. Plants' spatial life, defined against a background of quantitative landscapes, also becomes a practice of the number. In Clements's words: "Every plant is a measure of the conditions under which it grows."[77] Plants do not merely respond to the continuities and fluctuations of the quantified landscapes which surround them, but they are measures of their own conditions of existence *as well as* a proxy of the site they inhabit: the plant "is an index of soil and climate, and consequently an indicator of the behavior of other plants and of animals in the same spot."[78] We even find a hint of this proxy thinking in Nathaniel Ward's much earlier nineteenth-century meditation that highlighted plants as a form of remote sensing across time and space: "Almost every different region of the globe is characterized by peculiar forms of vegetation, dependent upon climatal differences; and thus a practiced botanical eye can, with certainty, in almost all cases, predict the capabilities of any previously unknown country, by an inspection of the plants which it produces."[79] Here the passage that started from Humboldt's investigation of the geographic distribution of plants through the phytogeographic methodologies of gridding and sampling ends now in plants embodying the numbers they are modeled with. As in Wiesner's case, in Clements's work, plant sensing also becomes a nonhuman practice of measuring.

The elemental ("natural") techniques of measuring and zoning that we discussed in chapter 2 might not be taken as cultural techniques in the traditional sense of "cultural." However, they need to be understood as a recursive part of the agricultural, ecological, and biological epistemes and methods of how so-called natural life is seen as producing fundamental divisions and differences in the nonrepresentational kind of matter that is full of signaling processes.[80] As a matter of fact, Clements's view of plants as living measurements of their surrounding conditions was not an anecdotal characterization of vegetal spatiality but a summarized presentation of the premise that grounded his practical book on plant indicators. There, the correlations between landscape types and plant formations articulated by his earlier work are presented under the epistemic figure of the indicator, following a practice of agricultural indicators that Clements traces in an opening section devoted to historical references. Through the figure of the indicator, then, the book presents specific and easily traceable vegetal formations as living signs of landscape parameters such as slope, altitude, fire, and animal presence, among many other factors and processes (see figure 6.4). The whole method becomes a form of reverse-engineering through the logic of the proxy that complements the earlier presented quadrat sampling method in the full arsenal of "datafication" of ecological surfaces. Here the idea of making "plants into quasi instruments for directly measuring the capacity for environments to sustain their growth" is a substitute for building specific technological instruments that would act like plants.[81] In other words, it meant that instead of making artificial instruments to produce "data that could be *correlated* to the life of plants," the aim was to perceive the whole environment as measured through and by the plant.[82]

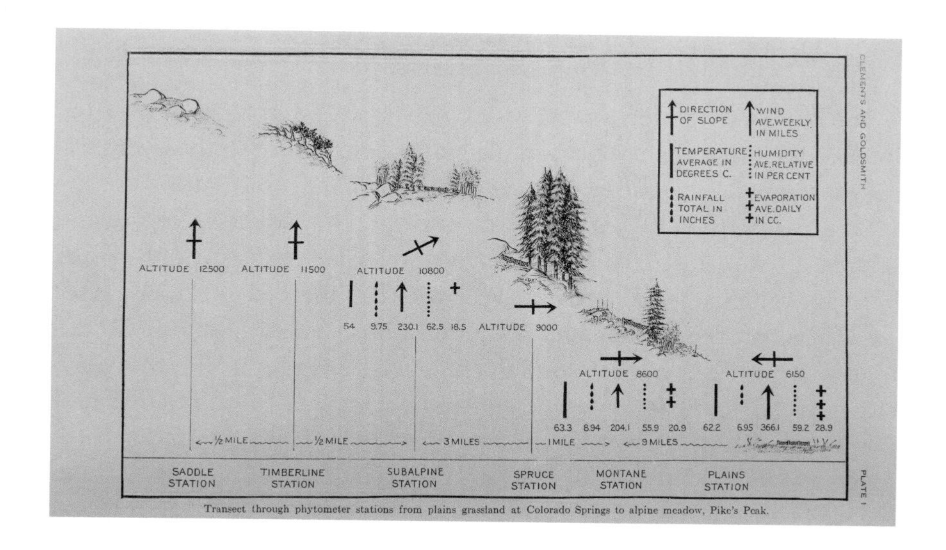

As we will see next, such proxy techniques and indicators did not stay limited to ecological research or agricultural use. As part of the weaponization of the capacities and characteristics of nature, this notion of plant indicators is directly related to remote sensing during and after the Cold War, in operational uses where the entwining of images and vegetal surfaces we are tracing becomes emphatically clear. In this way, Clements's analytical merger of plant sensing and zoning into a practice of measuring evolved into an "applied science" intelligence operation of sensing and a late satellite-era version of landscape physiognomy, reading geographical traits as signatures of their hidden insides.[83]

6.4

Transect from plains grassland to alpine meadow, featuring vegetation types as plant indicators. From Frederic E. Clements and Glenn W. Goldsmith's *The Phytometer Methodology in Ecology: The Plant and Community as Instruments* (Washington, DC: Carnegie Institution of Washington, 1924), plate 1.

PHOTOBOTANY AND COLD WAR PLANT INDICATORS

In the previous chapter, we discussed how the techniques for generating knowledge about the surfaces of the planet shifted during the twentieth century from the ground to the environments of images, resulting in complex infrastructures of hardware and software that could aggregate and address both terrestrial and extraterrestrial surfaces. On such platforms of images and data, once the appropriate circuit for remote sensing is set up and conveniently validated with ground truth data, it makes no difference whether or not the source of the images is physically accessible, either on earth or in outer space. "Digital twins" already exist in different institutional practices: these are environments of images and data (or images as data) that provide the necessary testing grounds for running relevant operational simulations.

Operational image layers multiply; simulations and other techniques give rise to image-based circuits that further transform the surfaces from which they emanate. These circuits also respond to what Orit Halpern signals as a characteristic of visualizations: "They make new relationships appear and produce new objects and spaces for action and speculation."[84] So far, our book has explored this ontogenetic character of data visualizations and images, including all sorts of targeting, defining, and refining surfaces through a myriad of tools and cultural techniques. And yet, surfaces are not passive even if taken as material for design in all sorts of ways and with all sorts of means that have implications far beyond the original agricultural connotation of cultural techniques. The meaningful relation that surfaces and living formations play as expressions of their own conditions of existence becomes a sign of what sort of *environment* of images they are. The continuous monitoring and observation of these surfaces are premised on the idea that they are already monitoring their own surroundings, acting as sensors and indicators. "Nature is a language, can't you read?," as The Smiths put it in the song "Ask." For the cybernetic-data period of remote sensing it could be rephrased as "Nature is a trace of patterns and signatures, can't you sense?"

For sensing to happen, it has to have happened already: photosynthesis, plant growth, chemical reactions from solar energy to soil ecologies, plant formations to the quasi-organisms of autogenic succession where plants shape their surroundings.[85] Technical sensing is, in this way, trying to catch up, to produce images of what has taken place, to establish categories such as vegetation indexes or engage in producing other aesthetic knowledge forms along the axis of seeing-to-classification. This recursive trait of *sensing sensing* characterizes a plethora of operations that form a genealogy of contemporary data culture, from labs, such as sensing human sensing in psychophysics, to, in our case, landscape surfaces of plant growth.[86] An extended and active surface that is not only the object but also the productive affordance of remote sensing is defined as a new environment of material relations: dynamic vegetal surfaces of growth and change, technical images, data circuits, and platform operations as coproducers of these relations that feature in different institutional practices.[87]

To address the nuances of this understanding of the environmental surface as affordances, we will discuss a case that took place during the Cold War, where the lack of physical access to the enemy's territories forced the intelligence agencies to deploy several initiatives geared at generating valuable new data out of aerial images that were also interpreted in relation to ground observations of plant-soil relations in different geographical areas. Methods from plant ecology became the mediator in this interpretation of plants as indicators and expressions of their own conditions of existence.

Consider this as a case of an abstracted yet active vegetal surface prepared for comparative operational purposes. Robert Gerard Pietrusko has analyzed the role that the observation of plants, trees, and other vegetal formations played for the cartographers of the US Department of Defense, who developed techniques of comparative aerial reconnaissance aimed at obtaining meaningful details of the Soviet landscape. The aim was to build models of the terrains based on the vegetation data obtained from high-altitude aerial photographs, while also actively mapping similar ground data that helped to analyze operational conditions, for example, for troop and vehicle access. With the help of the expert knowledge of botanists and ecologists, the teams sought to obtain plausible indicators of the slope, soil type, and soil moisture. As Pietrusko points out, these methods employed the ideas concerning plant indicators developed by Clements and others in the 1920s and 1930s.[88] Now though, the approach to plants as active surfaces of sensing and indication was not for "descriptive naturalistic terms" but because "plants directly indexed the ground's capacity to support military activities."[89] In short, the idea that plants became channels for gathering intelligence was not merely about their turning into media but into a very particular kind of operational media in the Cold War context.

The vegetation-covered areas were turned into indicator signatures, made possible through an exhaustive practice of comparing terrains and plant formations in the United States against the images from reconnaissance missions flown at high altitudes over the Soviet Union. Different expeditions sent to study ground conditions in North America provided the empirical data for detailed proxy analysis that could offer the speculative and comparative ground truth for visual analysis—for example, that "black spruce along the Albany River [indicated] an opportunity for camouflage, cover, and concealment," but also that the same vegetation could indicate "landscape's slope, soil type, and wetness" of significance for military logistics.[90] The observed and inferred correlations together suggested a series of plants and trees that could be recognized in aerial images. They provided visual clues but also invisual data for computational techniques of modeling. One could argue that they also produced the point about the vegetal cover as a data image in its own right. In Pietrusko's words, "Vegetation became a new form of photographic media":[91] vegetal formations surfaced as data sources feeding details of the terrains photographed so as to model an inaccessible region. The characteristics of the slopes that would eventually be crucial for the movements of troops and artillery in a conflict, for instance, were inferred from the systematic identification of vegetal species on the surface of the image.

One of the practitioners of this process of interpreting aerial photographs named this practice "photobotany," reflecting how aerial photography affected how the experts conducted their work at a distance.[92] First, it altered how they defined the plant communities that they studied. Pietrusko speaks of "photo-determined communities," where the team was interested in aerially imageable plants only.[93] Plant species that were small, sparse, or thrived in the shadows of other vegetation were not studied as they could not produce the proper spatial extent in these photographs. Second, the practice of photobotany also shaped the way the actual plants were characterized: "Their concern for vegetation's photographic signature affected the language that [was] used to describe the species."[94] Instead of the usual descriptors employed in botany and plant physiology, this analysis used a language grounded in tonal patterns, visual contrasts, and so forth: *a photomorphology* instead of a plant morphology, as Pietrusko emphasizes. This sort of "zoning" in and through the image provided a framework that built on the organicity of plant life while recursively treating it as data that could be appropriated into categories of operational use.[95] In this practical application of botanic indication, plants transform into light-emitting sources of signal intelligence, as photos transform into proxies of plant surfaces for morphological data analysis.

Pietrusko describes a case where the entwining of images and vegetal surfaces gave rise to a media-specific epistemic practice similar to those performed by the First World War RAF photo interpreters discussed by Paul Saint-Amour that we addressed in the previous chapter. Moreover, the case of photobotany makes it clear that there are material aspects of the imaged surfaces that enable the circulations to become meaningful for various epistemic practices, some more scientific, some more obviously operational. In particular, each specimen of the plant species identified by the experts was alive and maintained exchanges of matter, energy, and information with its surroundings when the aerial photographs were taken. This is not as obvious a fact as it may seem: many military surveying operations involved the extermination of vast extensions of vegetation, as we will see in more detail in the next chapter. Permanent features of much of the Cold War period were not only the weaponization of environmental data measurement but also the use of active agents—chemicals and pests—to kill crops and cause large-scale landscape damage.[96] In the case we are dealing with here, the actual life of plants was essential for the extended signal intelligence uses by the US Department of Defense. Plants as sensing biological entities were part of a broader circuit of capture and interpretation of different kinds of signals. This meant in this case understanding how the ecological exchanges between the plant, the soil, and other neighbor plants could be abstracted in the photointerpretation process and lead to operational knowledge of soil conditions. Just by being alive and connected to their surroundings, plants provide the elemental labor necessary for their operational functioning.

ADVANCED PLANT TECHNOLOGIES

Fast forward to 2017. The Defense Advanced Research Projects Agency (DARPA) announces its interest in plant studies, although on the more operative side as they declare the potential of Advanced Plant Technologies (APT) for military intelligence. One could claim that this is a mere continuation of what had already been in place since the Cold War, not just weaponizing different chemical or biological agents for active warfare but also developing techniques of interpretation and sensing that work through already active capacities of plants. The APT project presents itself as wanting to harness "plants' innate mechanisms for sensing and responding to environmental stimuli, extend that sensitivity to a range of signals of interest, and engineer discreet response mechanisms that can be remotely monitored using existing ground-, air-, or space-based hardware."[97] Unlike Clements's measurements, which were done by plants just being plants—after all, "plants are the best measure of plants"[98]—these plant sensors were already modified by way of the "advancement of technologies for performing multiple, complex modulations to plants, without sacrificing their environmental fitness."[99] Thus APT was conceived fundamentally as about genomic modification and synthetic plant biology, the central element of its promise being that nature was programmable: "Emerging molecular and modeling techniques may make it possible to reprogram these detection and reporting capabilities for a wide range of stimuli, which would not only open up new intelligence streams but also reduce the personnel risks and costs associated with traditional sensors."[100] The choice between designing artificial devices to act like plants or making plants act like (scientific) instruments was resolved by doing both.[101]

One would be tempted to see such large—and speculative—research projects as the culmination of the narrative we have presented, from plant ecology to the proxy type of indicating and sensing that informed Cold War period intelligence gathering. Plant landscapes could be read as proxies of their ecological formations and soil types, and interpreting images of those landscapes could be like reading plant morphology: a surface upon a surface that looped together in totalizing operations research that made cybernetics and ecosystem thinking seem modest by comparison.

Yet, even without such linear pathways of history that leave out many twists and turns of detail, it would be fair to say that APT crystallizes many elements of quantification of plant formations and demonstrates how their interpretation bends to many operational uses across the twentieth and twenty-first centuries. This case also presents a strong link to how environmental sensing starts as part of the narrative of the computational, programmable planet. As Jennifer Gabrys has shown, this view presents different scales of the earth as "an object of management and programmability,"[102] while creating technogeographies where "sensors as proxies are not standing in for a more-real version of environments, but are rather sensory operations that mobilize environments in distinct ways."[103] The reverse engineering at play is premised not to read from technics to nature but to place different combinations of technics-as-nature and nature-as-technics at the forefront of sensing networks, indicating, interpreting, and quantifying with a plethora of different institutional directions. Such fabrication of sensors and sensor networks is a version of the "managerial surface" we pointed to, which grows from quantitative management to management by creating sensor environments as vegetal surfaces. Contra Kittler, war and the military were not always necessarily the starting point but the end application of many technological and scientific innovations.

7

INTO THE FOREST: LIGHT AND RECURSIVE SENSING

Envelopes, globes, spheres, and enclosures; vegetal surfaces do not stick only onto a two-dimensional matrix but circulate across volumes of air, chemistry, and light. Images dive inside into the vegetal layers of the earth's critical zone. They infiltrate other realms of light, deepening the surface of observation and environmental modification. One prominent site of this dive, among others, is the forest, at the center of decades of discourses of planetary management and an even longer history of other cultural techniques.[1] The forest, at the center of this chapter, is a category that names a series of biomes that feature a stabilized milieu of exchanges and relations, where multiple species grow and compost under a familiar self-regulated environment of light created by the canopies of the trees and palms. To state the obvious, forests differ; they range from the northern coniferous zone to, for example, the rainforests closer to the equator. Different political histories meet with different cultural histories. Among the varied literature on forests, we want to emphasize two ideas that stand out concerning our core argument and approach. The first relates to how forests have become metonyms for the earth as a whole; they stand as paradigms of complex and integrated ecosystems and are, therefore, milieus for speculation about other ways of relating to the planet.[2] They have recently become understood as technologies in their own right as they are harnessed as part of mitigation strategies for planetary-scale environmental change. Forests act as carbon sinks, a fact that should not be treated separately from the histories of forests as spaces inhabited by indigenous peoples, for example. Afforestation projects can also be seen as large-scale terraforming related to multiple geographic politics of humans and nonhumans.[3] The second idea is political as much as it is aesthetic, as it relates to how the shadow world of forests has been linked to practices of hiding and fleeing, and thus as countertechniques to scorched earth, defoliation, and advanced vision technologies. A long history could be told of how military campaigns relate to forests and how forests have become large-scale visual media in and beyond military contexts. Forests are spaces of light and lack of light, of techniques of misperception as much as perception. At least in the Western political imaginary, the forest has been conceptualized politically as a site of freedom compared to other landscapes that are ruled as sites of serfdom and violence, including the violence of deforestation found in several sorts of plantation systems.[4] Forest fires signal another kind of violence on a planetary scale.

In this final chapter, we extend our discussion into transparency and opacity as they appear at the scale of a landscape. Surfaces are defined by their vegetational cover in complex ways that feed back into defining what visibility, light, and visual technologies of observation are. Beyond surveillance, this concerns the vegetal materiality of surfaces such as tree canopies and how these enable and disable particular views. In recent terms, this is further complicated by so-called AI technologies of visuality that extend, bend, transform, and recontextualize what is considered visual and visible. Building on the legacy of remote sensing, such data-driven observation of forests has become a key part of digital practices of "smart forests,"[5] as Jennifer Gabrys argues. As we will see, however, this does not point only to how such sensing and observation reinforces the idea of forests as resources to be managed. New forms of assemblages can be thought of, echoing the idea of medium design, where all sorts of human and other-than-human agencies participate. Paulo Tavares's work is an example of what type of result might come from this type of practice and research. While we develop this argument about forests and visual culture, we also end the chapter by a coda that summarizes many of the themes in relation to the argument about an ecological aesthetic.

IMAGES IN THE DOMAINS OF LIGHT

Forests are environments made of light. The vegetal specimens that make up a forest rely on light to grow a unique individual form. But, in addition, forests are made of light in the sense that walking into a forest means wandering into a luminous atmosphere clearly differentiated from the one found outside. While populated with diffuse and unclear signs of whether the forest advances or retreats, its frontier, the threshold, delimits two completely different environments.[6] In contrast to the daylight outside, light inside moves through the canopies as "a gigantic green cascade frozen in its downfall."[7] Light is projected and reflected in a slowly changing architectural assemblage of beams, shadows, and surfaces. The forest emerges as a space of shades and gradients, where green mixes with other seasonal colors, whether yellow, red, or blue. Entering a forest and suddenly being folded into its embrace is such a characteristic experience that also produces a specific mode of embodied perception. In this manner, there is also an affinity with the immersion in the architectural enclosure of images that we know as cinema: "In the beginning is the darkening. Indiscernible. Drifting into twilight. Attention gets lost in the space. Darkness comes back through the depths. Shapes and boundaries blur. Inside and outside are indistinguishable. Desert, void, blind land between sundown and circuits night. Like closing the eyes. Departing from oneself. Back to the beginning. And then radiating, shimmering, brightness, reflections, flickering. The trickling of light."[8]

The quote is the opening paragraph of Ute Holl's book on the cultural techniques linked to psychic processes that tie spectators to images in the movie theater. It describes the ritual of entering the temporal domain of the screen inside the space of projection. However, these lines could also describe the entrance into a dark forest after a morning of strolling under a bright sunny day: the threshold into the penumbra under the canopies and the accustoming of the eyes to the new light conditions. "The forest is a narcotic and intimate space, turning the inside out and bringing the outside in,"[9] as Matthew Fuller and Olga Goriunova put it, with words that resonate with those of Holl. They resonate, too, with the whole cultural history, folklore, and anthropological rites of the forest as a place where images are produced, such as dreamy apparitions, hallucinations, or even trance experiences.[10] In more modern terms of media, this could be called cinema as far as it is discerned by the cultural techniques of the modulation of light: the volume created by the light going through and behind the leaves and the radiance of the surface of cinema is also what comes to define a media archaeology of the arboreal space of a forest.

Such elemental media from forests to cinema can be described as part of a bundle of cultural techniques related to environments and projection architectures. According to Giuliana Bruno, writing about the "real ambient milieu" of projection, "It relies on a passage of light but encompasses other elements of a transitive, reflective, refractive, absorptive, unstable, and fluid experience of material space."[11] More specifically, Bruno addresses the environmental character of the effect of the projection of light in the film theater. Explaining that the screened images are less visual than environmental, she relates the radiant surface of the screen to the immersive articulation of light and space that features in the installations by James Turrell. In the artist's geometrical spaces saturated with light, "a viewing experience takes place and is configured spatially," Bruno writes.[12] Echoing Georges Didi-Huberman's observation about Turrell's chambers, this viewing experience occurs "not as a mere 'looking at' but rather as a 'looking into.'"[13] For Bruno, this is precisely the surface condition of the movie theater, the cinematic *looking into*: "the surface, turned into chamber, becomes habitable space."[14] Viewing becomes spatial through a "renewed 'camera' obscura," where psychic projection and architectural materiality coalesce in a semicontained space probed by flickering light.[15]

While in parts of this book we have extended a similar concept of surface to other scales and environments, such as landscapes and territories, we proceeded mostly by observing the recursion of cultural techniques operating at levels different from the psychic, the phenomenological, or the representational. In this section, we are interested in following Bruno's line of thought and extending it into the forest's interior as it becomes an aesthetic apparatus in its own right. Along these lines, the environments of light created in the interiors of forests echo this cinematic *looking into*, as it occurs in the work of the light artists analyzed by Bruno. When entering the forest, one is not only entering a space wherein to (eventually) get lost or come into potentially dangerous encounters but a place traditionally associated with visions of the enchanted: apparitions, ghosts, and other fantastical creatures. Forests are undoubtedly filled with the whims of shadow and light, which has inevitably given rise to hallucinated images of multiple kinds.

These whims of shadows and light, however, do not only feed the scared and playful imaginations of humans. A rich human and nonhuman world is incorporated in the forest, as Eduardo Kohn has shown us in his take on transspecies relationality. This anthropology beyond humans gives insights into worlds of meaning—such as in the tropical forest where Kohn's points emerge—fundamentally also *amplifying systems* for understanding how life as a distributed system thinks. Semiosis is not restricted to humans.[16] In this line, the imaging force of the forest, too, can be witnessed in a projective surface of images of a very different kind beyond the dread and delight of a singular subject. On a sunny day, for instance, when light moves through the canopies, some of the rays may travel uninterrupted between the leaves and reach the ground, sprinkling its surface with flecks of the sun. Forest floors under dense canopies emerge as screens scattered with patches of light. Some are large, some are small, but all of them display a similar elliptical shape as the sunlight filters through the leaves in a series of tiny, overlapping disks on the ground, different from the larger sun flecks that appear under lighter canopies:

> The surprising thing is that all these spots have the same shape, and yet it is unlikely that all those chinks and openings should happen to be so nicely similar and round! Intercept one of these images by a piece of paper, held at right angles to the rays, and you will see that it is no longer elliptical, but circular. Raise the paper higher and the spot grows smaller and smaller. So we conclude that the pencils of light forming such a spot have the shape of a cone and the spots are elliptical only because the ground cuts this cone slantwise.[17]

The overlapping disks under dense foliage are round because they are literally images: they are pictures of the sun (figure 7.1). Sunlight goes through the holes in the canopies as if through the pinhole of a camera obscura: a cone of light is produced. When it meets the ground, the disk of the sun outside becomes an image inside. It is a double inversion, additionally, not just bringing the outside to the inside: "Just as if each minute, leafy aperture were actually a lens or an object glass,"[18] the images of the

7.1A

Forests as cinematic environments of light. From Mabel Loomis Todd, *Total Eclipses of the Sun* (Boston, MA: Roberts Brothers, 1894), 20.

7.1B

Forests as cinematic environments of light. From Amédée Guillemin, *La lumière* (Paris: Hachette, 1882), 41.

sun look upside-down, which is a fact that can be easily confirmed during a solar eclipse when the optical phenomenon is most wondrous.

Cinematic viewing in the forest becomes environmental through a renewed camera obscura where psychic projection and arboreal materiality coalesce. We are here again echoing Bruno but also extending her work on atmospheres and surfaces to encompass the broader vegetal landscapes as environmental media in their own right. As it occurs in the movie theater and in the above-mentioned light installations, the forest exists in an equilibrium sustained by interweaving material surfaces and light projections. Viewing becomes environmental, and looking becomes a *looking into* because a particular modulation of light conditions all the elements and living creatures inside the forest under the canopies. There, light is not only an index of an inside but a space for circulation, where images, signals, and surfaces meet. It starts to become thus also cinema without human viewers: a material transformation that is inscribed as informational traces into the trees, soil, and other organisms.[19] The forest is an already existing sensorial mechanism that records its own data and gives insights into "aesthetics beyond perception."[20]

LIGHT CLIMATES AND PHOTOGRAPHIC COMPUTATION

In a 1974 article in the *Journal of Applied Ecology* provocatively entitled "Forest Cover as a Solar Camera," the authors noticed that even if pinhole images of the disk of the sun under dense canopies were a common phenomenon, no photographs of them had been published in scientific literature.[21] Their text consisted of a review of models of interception of light by canopies, together with photographs of the sun flecks and guidelines for those interested in learning the technique. The images were presented for reference and for teaching purposes, in this way echoing the role of photography in science education that we observed in chapter 2.[22] The importance of pinhole images—the authors emphasized—is that they are an extreme example of penumbral effects. Forest penumbra and, more broadly, *light habitats* are an object of scientific inquiry. Light habitats are zoned into categories such as forest shade, woodland shade, small gaps, and large gaps, milieus with associated surface patterns in the light exchanges and color signals between plants, animals, and insects: "Forests present a complex, changing, and heterogeneous light environment. In them, animals and plants signal visually to their mates, pollinators, and dispersers, and try to avoid sending the wrong signals to their predators, parasitoids, nectar/pollen robbers, and herbivores. Visual signals are a product of signaler's reflectance spectrum and ambient light and so are strongly dependent upon the environment. If the ambient light spectrum changes, a color pattern may change from inconspicuous to conspicuous or vice versa."[23] At the back of this notion of light habitat lies an assumption of a deep entwining of the ensemble of living creatures with the levels of light inside the forest. As a matter of fact, light habitats are also addressed as *light climates*, a term that brings to mind the broader sense that climate has when it is considered in relation to its influence on life forms, even on human culture and social institutions.[24] In her take on air as elemental media, Eva Horn remarks how this broader notion of climate—beyond the concept of average weather—indicates a locality, that is, a sense of place on a spectrum of difference that allows one to relate one place to another.[25] Thus the concept of light climate would not only define environments of light but also establish a ground for comparison of different instances. This was one of the key questions pursued by light climate research.

In plant ecology, this question of comparison functions through techniques of measuring the environments of light created by forests. If light climates can be measured quantitatively, then they can also be easily compared. Along these lines, during the first decades of the Cold War, as part of the political remaking of forests that facilitated the transfer of technologies to forestry departments in the United States, this problem was addressed under the rubric of scientific forestry.[26] With links to military operations, state-driven policies, and migration control, forests were scrutinized anew between 1950 and 1970. As a consequence of the rising interest in forest volumes, new methods were devised to measure and characterize their different light climates.[27]

The magnitude that the researchers sought to measure then was the amount of light that characterizes the interior volume of a forest. An equilibrium between the species of plants, the features of the soil, the overlapping of canopies, and the other multiple circuits of living creatures—eating, seeding, pollinating, or destroying it—characterizes a woodland. Do the different types of forest equilibria correspond to different values of the average intensity of light inside? In the context of this book, the question is not new. It is related to the same domain of investigations formulated by Julius Wiesner when he inquired about the average intensity of light value needed for each plant species to grow. As we showed in chapter 2, he named this value the plant's Lichtgenuss, its appetite for light. Not by chance, Wiesner's work and his Lichtgenuss are cited and commented on in the literature on light climates.[28] Lichtgenuss, nevertheless, was about individual appetites for light and light climates were about the collective ensemble of a forest. In this regard, it is worth observing that while a photometer comes to mind as the most adequate instrument for the task of quantifying a light climate, the wide fluctuations in the light intensity inside woodlands require the development of more complex apparatus. Even so, despite the differences between Wiesner's research and the problem of light climates, both cases ended up with a practical solution found in the domain of photography. While Wiesner forged his concept through his expertise in the sensitivity of photographic paper to light, the botanists and plant ecologists of the first decades of the Cold War found in hemispherical photography the tool to register and quantify forest canopies. In other words, once again, images mediated the ecologies of light and plants.

Before arriving at hemispheric photography, the photomosaic was one of the photographic techniques explored for capturing forest canopies and light regimes. With a standard plate camera, researchers took a large number of exposures and fitted the photographs together to form a hollow hemisphere. Through a specially designed projector, the hemisphere was projected on to a plane. Other approaches involved a pinhole device named a "hemispherical forest photocanopymeter," which, although it allowed for the capture of a 180° panoramic of the canopies above, resulted in poor definition, among other problems. The most apt solution for the problem was found in the context of weather research. The fish-eye camera, invented and constructed in 1924 for cloud observations, was proposed and explored in this context. The "Robin Hill Camera," so called after the name of its creator, who facilitated its adaptation to these research uses, resolved most of the difficulties of canopy photography, picturing the whole hemisphere on a single flat photograph (figure 7.2).[29]

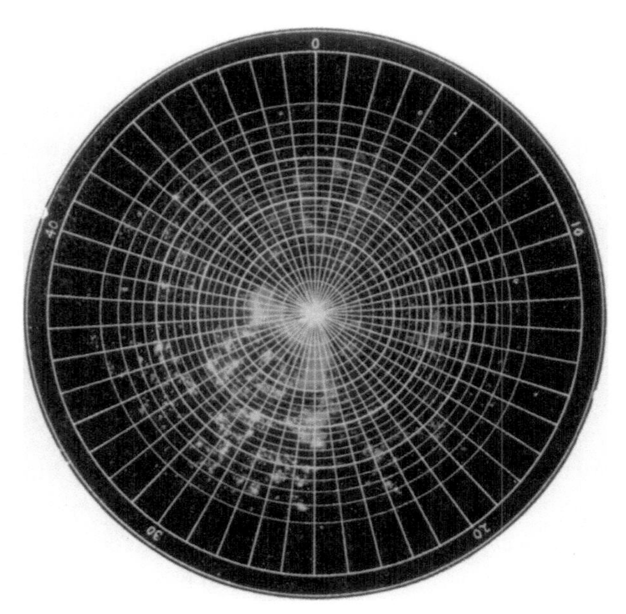

PHOT. 1. Hemispherical photograph of the small clearing in summer, with the grid for estimating diffuse site factors superimposed. Each segment of the grid corresponds to 0·1 % of the total illuminance from a standard overcast sky.

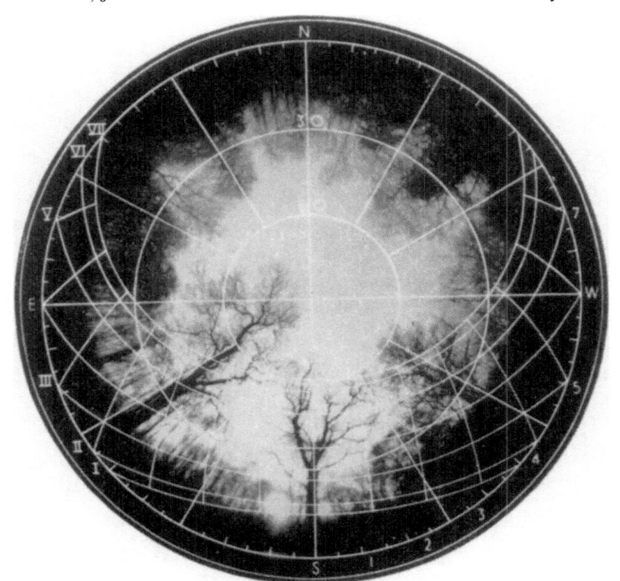

PHOT. 2. The large clearing in winter, with a superimposed solar track diagram for 52° 13′ N. Details of canopy structure to the south have been lost through the extreme brightness of the area round the sun, which, from its position, shows that the photograph was taken around 11 a.m. in early January or late November (it was actually taken on 10 January 1962).

(*Facing p.* 34)

As was the case with the aerial images linked to sensor measurements discussed in chapter 5, flattening the canopies onto the surface of a photograph allowed for the superimposition of additional relevant data in a new layer. This included, for example, solar track information or a polar grid of concentric annuli and radiated lines serving as an address system inside the image. Furthermore, grids became not only a layer for locating areas in the photograph but devices to compute the average reduction of daylight illuminance inside the forest.[30] Botanist Margaret C. Anderson called this method "photographic computation," explaining how it worked: "Having superimposed this grid on a hemispherical photograph of, say, a woodland canopy it is possible to estimate the percentage reduction of illuminance by counting the segments clear and those obstructed."[31] Grids allowed the segments of the image to be counted and classified: the techniques of enumeration of living vegetal surfaces featured in Clements's work on plant geography were used again in order to calculate an estimate of the light transmittance of the canopy. With the mediation of hemispherical photography, light climates could be documented, computed, and compared with each other, thus establishing a circuit of operational techniques of analysis. Even prediction was included in this computational operation of reading of photographic surfaces: "It should be possible to predict monthly figures for total, direct and diffuse light at a site from photographs."[32] Furthermore, these techniques enabled the remote characterization of geographically distant light regimes through photography.

With the spread of personal computers and smartphones, Anderson's computational grids have given way to software applications and

7.2

An example of hemispherical photography of a clearing in summer and winter, featuring Anderson's photographic computational grid (top) and solar track information (bottom). From Margaret C. Anderson, "Studies of the Woodland Light Climate: I. The Photographic Computation of Light Conditions," *Journal of Ecology* 52, no. 1 (1964), plate 1. Used with permission.

hardware devices such as the HemiView system or low-cost methods with smartphone fish-eye lenses to obtain a standardized calculation of the light transmittance of a forest.[33] Photographic computation has thus become part of the circulation not just of images but of easy-to-access techniques that extend the projective imaging force of the forest as an apparatus to digital environments of imaging. Not by chance, even imaging techniques such as LiDAR extend these forms of machinic vision and computation of the arboreal to the architectures of exhibition rooms and movie theaters.[34] Light climates thus come to exist as a special case of the immersive aspiration of the digital.

CLOUD IMAGES AND FLASHED SUNS

In the previous chapter, we saw how techniques developed in plant ecology became repurposed during the Cold War as they were turned into tools and resources suitable for military and surveillance operations. In this chapter, we have already noted how, also during the Cold War period, new techniques were developed to address forests as spaces whose opacity to state, border, and military control needed to be countered through newer means of monitoring and observation. Was the research on light climates also related to military contexts? Was it part of the infrastructural regimes of signal intelligence, an early example of remote sensing, and "operations other than war"?[35] In order to address this question, we will go back for a moment to the 1920s, when Robin Hill developed and published his invention of the fish-eye camera designed to produce whole-sky photographs for cloud observation.[36]

While Hill was occupied with cameras and hemispherical lenses, one of the major laboratories devoted to photographic research in the world at that time, the Eastman Kodak Research Laboratories, was engaged in a different type of project that involved the sky, photography, and, to some extent, weather too. The research, led by George W. Goddard, pursued a method to take aerial photographs at night. The experiences gathered during the First World War highlighted the importance of dark hours in relation to monitoring the enemy's movements and constructions, so techniques for night photography needed to be invented.[37] As a result of the project, on November 21, 1925, the first flash bomb was dropped from an airplane over the city of Rochester in New York. The falling cartridge—also called a photoflash bomb—consisted of a wooden casing filled with a special flashlight powder and a delayed-firing mechanism to ignite the container. Its explosion produced a peak of 100 million candelas of intensity of light—the equivalent of 10 million standard lanterns.[38] For a twenty-fifth of a second, the city was flooded with light coming from an artificial star (figure 7.3.) "All along the streets into town, there were groups of people standing around trying to figure out the cause of the terrific explosion and flash which had shaken the city," wrote Goddard.[39] A camera onboard, synchronized to the firing mechanism of the bomb, took a photograph. It was the first of many aerial night photographs to be taken, representing an early case of active sensor technologies that created the conditions of visibility for their own purposes. Since then, and before the advent of satellite imagery and nighttime optics, let alone such imaging practices of light pulses as LiDAR, photoflash cartridge design evolved while the bombs were used extensively as a source of light for aerial night photography (figure 7.4).[40]

If Robin Hill's camera was designed for weather observation, Goddard's flash bombs—and their intervention in the environments of light in which war took place—operated instead in the domain of weather transformation. In this manner, their operational use belongs to the domain of atmospheric politics that have characterized the twentieth century that brought with it the environment into the battlefield, following Peter Sloterdijk's elaboration of the concept of atmoterrorism.[41] Like the early cases in this book that concerned atmospheric envelopes as explicating the air and its domain, here, an awareness of not just chemistry but light as part of envelopes of visibility became a central trait of "enlightenment" (as in, concrete light). Photoflash bombs were able to suspend the darkness of the night and replace it with a spectral aura of instantaneous daylight for at least the time the shutters of the cameras in the airplanes remained open. Momentarily, they transformed night into day. Thus, the simultaneous development of hemispherical photography—devised for the observation of clouds against the whole sky—and the manufacture of

a system for lighting up the night sky with instantaneous artificial suns recalls again the two-way relation between observation and transformation that we have discussed continually during the book. This two-way relation is the double bind between images and landscapes we are tracing from glass (chapter 1) to grass (chapter 6). We have discussed, for instance, how atmospheric envelopes can be witnessed in the chemical seeding of vegetal growth (chapter 3). In this regard, we will now switch to another form of seeding, the chemical seeding of clouds. This way, we will see more clearly the double bind in relation to forests' light climates and will

7.3

A photoflash bomb detonates over La Spezia, Italy, during an air-raid on the night of April 13–14, 1943. It has illuminated the town's dockyard and a berthed battleship (marked "A"). The silhouette of one of the attacking Avro Lancaster bombers can be seen. Source: Imperial War Museum, catalog number C 3697. ©IWM. Used with permission.

NIGHT FLASH PHOTOGRAPHY

NIGHT flash photography is not actually new. Work on this method of illuminating miles of territory and photographing that territory from altitudes up to 10,000 feet, began in 1928 and was perfected right before the war.

The diagram on this page illustrates the angle at which the flash bomb falls. The flash is released to explode at a predetermined elevation. At the peak intensity of the flash, the camera shutter trips automatically, thereby photographing the area illuminated by the flash bomb.

FLASH BOMBS and a night camera are loaded aboard a plane for after-dark photography. *U. S. Air Forces.*

THE BOMB is released to explode at a predetermined height, tripping shutter automatically. *U. S. Air Forces.*

ERIAL SHOT taken at night with flash bomb. The area within the white rectangle reveals concentration of railroad cars.

7.4

Aerial night photography explained in the US magazine *Camera Comics* 4 (1945), 34. Source: https://archive.org/details/CameraComics004C2cJvj-narfstar1945/page/n51/mode/2up. Public domain.

elaborate further on the relationship with the military and the context of the Cold War.

WEATHER WARFARE

Research on television technologies brought new approaches to the problem of night vision during the 1950s. If low-light television images could be amplified when transmitted between screens, should televisual light amplification also be possible, enhancing the capacity to see scenes in the darkness? Due to this research, night vision techniques of the US Army during the Vietnam War did not rely on the instantaneous artificial stars blown by the photoflash cartridges but on the Starlight scope attached to the rifles of soldiers installed on helicopters and on fixed-wing aircraft.[42] They were electric light intensifiers, "a television tube of sorts," which rendered visible what the naked eye would otherwise barely differentiate from background darkness in and out of the forests.[43]

Photoflash cartridges were not left in the corner of a warehouse, however. Their use experienced an unexpected twist during the Vietnam War, as they were transformed into weather warfare. While from the point of view of military technology, the Vietnam War is mostly known for the use of chemical agents for the defoliation of green areas and for the cybernetic approach to the war, an experimental program of weather warfare was developed and put into practice on the battlefield. In particular, responding to what has been described as an atmospheric disposition that "sought to enclose, police, and pacify hostile forms of life," US aircraft released photoflash cartridges inside clouds to trigger the release of moisture.[44] Instead of flashlight powder, the photoflash cartridges spread in the sky particles of a chemical mixture that, once ignited, produced the silver iodine (photosensitive compound used in photography) to seed the clouds and make them precipitate. The same hardware whose bursts of light enabled aerial night photography was explored to envelop the photographed territories with controlled climate conditions (figure 7.5).

These techniques of weather warfare were not tried only in Vietnam. As Yuriko Furuhata writes, cloud seeding experiments had taken place in India on behalf of a purportedly humanitarian project of alleviating

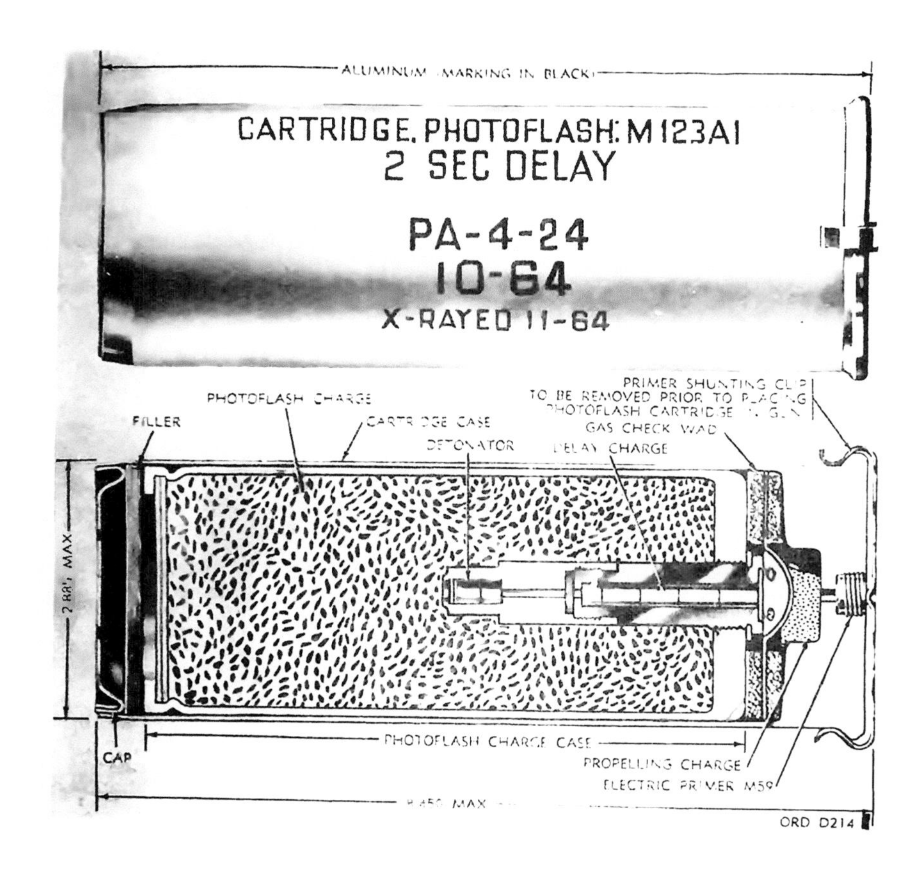

7.5

Illustration of a photoflash cartridge, showing the chemical charge and the delayed detonator. From *Department of the Army Technical Manual TM 9-1385-51: Identification of Ammunition (Conventional) for Explosive Ordnance Disposal* (Washington, DC: Headquarters, Department of the US Army, 1967), 72.12.

droughts.[45] Later on, a scientifically controlled test of the photoflash cartridges was conducted over a North Vietnam supply line in the Laos Panhandle, where over 85 percent of the clouds tested reacted favorably. The conclusion was that cloud seeding could be used as a valuable tactical weapon. Weather reconnaissance aircraft and photographic reconnaissance airplanes were used to seed the clouds, the latter carrying seeding units in the photo cartridge compartments and dropping them inside the clouds at intervals of approximately one kilometer, influencing an average of four or five clouds or groups of clouds per day.[46] It is not clear, however, if the operative phase of the cloud-seeding program and its expansion to North Vietnam and Cambodia were effective, as not enough information was gathered, in particular about nonseeded control clouds.[47] In any case, as Ian G. R. Shaw has put it, "industrial-scale atmospheric power had arrived in Southeast Asia."[48]

Weather warfare sought to extend the effects of the rainy season and transform the ground conditions accordingly: to flood the soils with water, to fill the atmosphere with humidity, and to distort vision and affect bodies with fog and moisture. It would seem then to go against the pervasive monitoring layer of the war that the Pentagon tried to set up, where among other things, "there was to be no 'fog of war.'"[49] This contradiction, however, is only apparent: the weather warfare can be understood as part of the Pentagon's aspiration to reengineer the physical atmospheres of Vietnam into an abstract landscape that could be efficiently measured, addressed, and bombed.[50] Certainly, the primary aim of the project was logistical—to create poor traffic conditions by removing road surfaces through excess water and landslides—but as Furuhata has argued about geoengineering, the context of its emergence entails a "technophilic desire to posit atmosphere itself as an object of calibration, control, and engineering."[51] That is, the muddy ground and moisture after the seeded monsoons was part of a project to expand the addressable, computational, and predictive space of the grid to the turbulent accumulation of surfaces and the transformed atmospheres of targeted locations, including the ground conditions of the forests and other vegetal formations.

The flash bombs turned into weather warfare bring Giuliana Bruno's discussion about the spatial character of light projections onto a different scale. Here, we see skies becoming screens of light and atmospheres, in the literal sense, being modified by the same means. Seen from this point of view, the concept of light climate does not sit only between vegetal sensitiveness and the molding force of light, as it did with the related notion of Lichtgenuss. Neither does it sit only between the imperceptibility of smooth changes and their statistical accountability, as in the case of the prairie surveys. It fits in the juncture between the observation of environments at large as coherent envelopes and the techniques of modeling, modifying, and forging such spheres of dynamic atmospheric equilibria.

The above-mentioned idea of forests becoming metonyms for the entire planet echoes again, but in a different sense. Margaret Anderson's method for the photographic computation of light climates combined the camera for cloud observation and its hemispherical lens with the polar grid for numerical calculation. The cloud camera was brought into the forest, transforming the canopies into a celestial dome of sorts as if the image were projected onto a virtual sphere. The resulting images recall the "dreamland portraits" reflected on "the evanescent surfaces of a simple soap-bubble" that experimenters such as Charles Vernon Boys photographed at the beginning of the twentieth century, linked to the glass enclosures we discussed in chapter 1.[52] Inside the forest, the canopies surrounding the hemispherical camera seem to diagram an inversion of the sealed glass envelopes where plants had made their mark as creators of livable atmospheres. The uncanny projection of the otherwise familiar embrace of the trees seems to render, in visual terms, this nesting of scales. This suggests a peculiar hypothesis: the environment would not reside in the forest except contained in the sensor sensing the climates of light—like the cloud-seeding powder contained inside the flash bomb cartridges.

Anderson's method already embodied the merger of climatic conditions and vegetal surface calculation that would, decades later, characterize the development of the earth system models (ESMs) that anticipate the current digital twins of the earth. These conflations are fundamentally to be seen as recursive operative chains of models that do not lose sight of the real but that shift their terrain of influence to second-order cultural techniques, medianatures of sorts. As Paul N. Edwards has shown, the dominance of numerical weather prediction projects in terms of resources and financing during the second half of the twentieth century set the ground for the adoption of climate modeling standards in ESM simulations. There, atmospheric and ocean circulation is analyzed in relation to other layers, such as vegetation. That was the case, for instance, of the development of the equilibrium vegetation ecology model (EVE), a vegetal landcover component that was part of the first examples of these models.[53] Following Edwards, the EVE component relied on a grid resolution different from the ones used traditionally in global biome classifications (too broad) and in plant ecology surveys (too narrow). This resulted in the formulation of a medium-granularity "life form" concept that abstracted vegetal surfaces to become a compatible layer in the spherical simulation environment.[54] In this regard, the introduction of Margaret Anderson's hemispherical globe of photographic computation into the forest resonates with the later standards of global climate modeling, as they introduced their own globular grids to compute the canopies from the scale of the planet as a whole. The play of scales of living surfaces echoes here once again, as do techniques that help to move across scales by way of images and models in relation to landscapes, which themselves, as we quoted Jane Hutton in the introduction, "are models *in situ*."[55]

A MULTISCALAR PLANETARITY

So far in this book, we have discussed the entanglement of media and cultural techniques with terrestrial surfaces we refer to as the environmental living surfaces, by focusing on how they spread at different scales. That has been the case in the chapters focused on vegetal leaves, agricultural plots, prairies, the biosphere, and even other planets as well. Continuing with this dynamic of spreading, we could have analyzed forests from this perspective and approached them as a receding type of vegetal formation in favor of the surfaces of cleared visibility that have historically replaced them due to the threat that their opacity has posed to different forms of power.[56] Scorched earth techniques such as burning forests against insurgent guerrillas or harming fleeing populations would be examples of this, as are forest fires in another register of ecocide. Incidentally, the cases discussed in the chapter are related to the efforts by the US Army in the Vietnam War to expose a "hidden enemy" under the thick vegetation of the jungle.[57] Multiple layers of militarization of vegetation were at play in Southeast Asia, with the US-led defoliant and herbicide strikes under Operation Trail Dust and Operation Ranch Hand relying on the infamous Agent Orange. This dynamic of receding forests as visual ecocides continues to the present, as Hannah Meszaros-Martin has made explicit in her research about the use of fumigation chemicals over tropical forests as an environmentally critical component of the visual surveillance from above in the antidrug war in Colombia.[58]

Our focus on the concept of light climates, however, entails a different point of view in relation to forests. Light climates and their quantitative analysis through hemispheric photography have prompted us to go under the canopies and focus on the liminal space between opacity and transparency that characterizes their arboreal interiors. This opens up the domain of practices of images that unfold inside forests, together with the shadows, reflections, sun pictures, and other active registrations of events of all sorts. Something similar also characterized the visual operations in the midst of canopies of Vietnam during the war. Not only to see forests but to see through forests and to be able to construct images that layered significant composite information became part of the spatial and architectural image complex that mapped potential operational data.[59] Different forms of technical sensors during US Operation Igloo White were to provide a signal link from the ground level (including under the canopies) to aerial surveillance, and in general, different ways of relating to the forest ecology—according to Pujita Guha—marked bespoke environmental attunements that, to paraphrase Guha, modulated the relations between sensors, sensing, logistical movement, and different vegetal milieus.[60] In this regard, such sensor arrangements represent an early version of imaginaries of environmental management through remote sensing, as Ryan Bishop has put it.[61] As a matter of fact, they have evolved to come to define a domain of *precision forestry* similar to that of agriculture.

Below the canopies, however, the arboreal and the data cultures of sensing and automation might fold into more complex entanglements than the Cold War imaginary of total monitoring (which never was, of course, *total* and instead prone to persistent errors). As Jennifer Gabrys writes, smart forests form environments where assemblages of living and nonliving matter are enmeshed in networks of processes of human and nonhuman agents that ultimately redefine what forests are and what they can become.[62] Sensors and other technologies are not introduced into the forest as if it were a passive background to be measured and sampled but as a media environment where light, carbon dioxide, heat, and other forms of matter and energy are registered and expanded as events. Beyond sensing technologies, Gabrys's research takes into account the action of automated processes—such as fire prevention techniques—and the effects of civic technologies designed for community participation and engagement. In this regard, forests are not taken as only generators of environmental data, nor as entities individuated and separated through technoecological regimes of regulation, but they are themselves recursively becoming sociopolitical technologies for addressing phenomena of a different scale, such as environmental change. In Gabrys's words:

> Technologies that would govern these environments in order to address and mitigate environmental change are also transforming them into entities that are meant to operate as technologies. Smart forest technologies and systems thus contribute to the reworking of what a forest is and how it operates. Such a point of orientation could expand existing approaches to smart and digital technologies, while reworking conventional understandings of what counts as a technology, by researching how forests transform into technologies through technical and policy interventions.[63]

Our book tracks similar recursions that have appeared, for instance, in the form of plant indicators for military environmental intelligence, which was the topic of the previous chapter. Here, however, the recursive techniques also entail the circulation of the bodies, practices, and technologies that make up the expanded networks of sensing, seeding, maintaining, computing, and other forms of relation with the forest. According to Gabrys, the surface tensions under the canopies make them become not only metonyms of planetary equilibria but also experimental assemblages where forms of planetarity can be assayed distinct from the globular grids and their embodied legacies of the Cold War.[64] Experimentation is often subsumed into corporate modes of environmental management, but a focus on experimentation also recalls Matthew Fuller and Olga Goriunova's take on the forest: "Living in proximity to the forest, we propose, means living in a thoughtful and experimental relationship to matter and to movements of energy of life," where one is prompted to engage "in material acts of experimentation that accrete as technologies."[65] Elaborate techniques of sensing and living are already embedded in forests.

These technologies and acts of experimentation involve a process of defamiliarization, an openness to the unhomely when facing the radical otherness of the planet involved in Gayatri Chakravorty Spivak's concept of planetarity.[66] Alongside Spivak and Gabrys, Fuller and Goriunova acknowledge this as the uncanny aspect of ecological aesthetics: "Things become uncanny when scales or conditions that are normally unrelated to a matter or to a problem are drawn in as a means of negotiating a situation and which thus cast them in a new light."[67] In this regard, the episodes of the genealogy of the living surface traced in this book have shown how images have cast a new light on vegetal phenomena. At very different scales, we have seen how methods that one would expect as "normally unrelated to [vegetal] matter," paraphrasing Fuller and Goriunova, have brought unexpected associations between the photosynthetic and other registration of light. These have taken place mostly inside the colonial, technoscientific, and military-industrial complexes, producing techniques of management that operate through the surface. However, alternative counterpractices emerge in the midst of this combination of advanced imaging and forest surfaces.

This is the case in the work of architect and researcher Paulo Tavares. His research has shown how imaging techniques such as radar-based cameras have transformed the Amazon rainforests into a deeply monitored space. These techniques have evolved in the last decades into algorithmic-driven multifrequency radar scans, which, depending on the frequencies used, allow us to see either below the canopies or even below the ground (as in ground penetrating radar).[68] Imaging techniques render the canopies transparent for extractive industries such as mining, echoing points made by Geoff Manaugh we referred to in the introduction.[69] Such geomedia produce biophysical and geological inventories as part of the view of forests as resources. However, the same images and imaging techniques have brought to light traces of human practices that are unknown and incompatible with the Western conceptualization of the Amazon. LiDAR data, radar scans, and the open clearances evidence large-scale geoglyphs as well as leveling practices below the canopies, which have also been linked to patterns of arboreal distributions across the forest. In Tavares's words, what these images cast is that "the Earth's largest biodiversity refuge is to a great extent a heritage of indigenous design."[70] In one move, the idea of the Amazon as a form of pristine nature without humans is debunked, while the very notion of design is redesigned:

> New evidentiary technologies—from remote sensing large-scale environmental transformations to the micro-forensic analysis of fossil seeds—are allowing to make visible the many different forms by which the forest, as ecologist William Balée writes, configures a great archaeological archive that "harbors inscriptions, stories and memories in the living vegetation itself." In the same way we read the city as a historical text produced by social forces coded into material form—layers on top of layers of ruins forming a living social fabric—the forest stands to be interpreted through the syntax of spatial designs. Yet these living ruins are neither fully or exclusively human nor completely natural. Rather, they are the product of long-term and complex interactions between human collectives, environmental forces and the agency of other species, themselves actors in the historical process of "designing the forest."[71]

The logic of the map as a cultural technique of design is inverted in Tavares's reading of the indigenous urban forestry of the Amazons. Maps do not surface here as modern future-oriented navigation instruments, as media that defines space from an exterior point of view. In this type of premodern design, every arboreal volume is taken as an entity already modified through human and nonhuman habitation and modified as such. Forests are observed through this recursion, where mapping emerges as dynamic interaction practiced on top of a background of ongoing correlations.[72] Consequently, forests are modified as a self-preserved future form of the ruin itself, or living ruins, in Tavares's words. They are images of the past, but they are also images of the future, past tense. Here, the living surface emerges inside the forest as a constitutive principle and a navigational tool, an expression of what can be called elemental media and even medium design, in the sense of tracking potentials and dispositions: if design relates to tracking "interdependent components,"[73] some of these are also found in historical, even archaeological layers. Under the canopies, recursion is materialized as the trace of the bygone and the yet to be.[74]

CODA

Every new futurism has its own kind of surface. Depending on the period and the aims, they have ranged from glass to plastic, vegetal plantations to data. The current fascination with the vegetal, which Rob Nixon locates in the idea that "the forest serves as biological precedent and loose allegory for a shared survival from which the self cannot be disentangled,"[75] has an unexpected development in forest plantations around the world becoming bio-tech industries: forests made up of trees and plants designed and synthesized to autonomously grow biotechnological products in the open air, as a photosynthetic industrial complex of sorts.[76] The reality, though, might turn out to be more modest than a total new resurfacing of the entire planet. However, tracking the variety of imaginaries along surface sensoria proves a useful task for a cartography of the planetary as a multiscalar patchwork of concerns.[77]

The urban and technoscientific cultures of glass did not give rise to the forests of glass tubes nor to the gardens of glass flowers that featured in the early twentieth-century fantasies of Giacomo Ciamician or Paul Scheerbart (chapter 1). However, glass cases contributed to the spread of the plantations that retextured vast expanses of territories linked to colonial trade. In the recursive loop discussed at the beginning of this book, nested glass became a model of the cells that make up vegetal green matter. Plant physiology integrated the glass perspective in its repertoire of techniques and views. Glass enclosed and separated environments in order to transport living plants. In an interplay of insides and outsides, the glass sphere enabled a transformation of the said environment too.

The environments were sealed, but light could travel through them. With light, other things could pass too as interscalar vehicles: images, measured intensities, and other forms of data transmitted through electromagnetic waves.[78] The realization was that all these forms of light belong to the domain of energy exchanges that only plants, bacteria, and algae among the living can process. As a result, the sensitivity of vegetal cells to light was described in photometric terms. The growth of vegetal surfaces was visually captured, in turn, with the help of cultural techniques such as time-lapse photography. This entwining we describe has gradually arrived at the point where the cultivation of plants, in the context of digital farming, has become a matter of circuits of data images such as those of precision agriculture. Architectures of experimentation with light, such as phytotrons, are matched by the industrial hydroponic farms of artificial light and agriculture. The cybernetic modeling of ecosystems from the perspective of flows of energy gave rise to a broader push toward environmental management, while it also featured in many concrete architectural plans for enclosures of the living, such as the "cabin ecologies" of the period of the Cold War space race, just one of the many examples of biospheres discovered, imagined, and designed across the twentieth century.[79] Invertedly, this also fed back to how existing environments, such as the planetary biosphere, came to be seen as one large case of such cabins. In this persisting vision, a new kind of surface—defined by the managerial interweaving of plants, soil, and images—envelopes the world. This envelope is not just a stable cover, but is as much about logistics of movement, returning us to the opening scenes of the book: the green mantle of the planet is actually a mantle on the move as different materials, flows, and samples are uprooted, transported, replanted—and repeat.

As we have seen, the visual circuits of digital agriculture can be linked to earlier apparatuses of state and corporate powers that, through aerial photography, engulfed and regulated not only the plants' metabolism of light but also wider spatial and temporal domains of infrastructure, logistics, and labor. The relation to the inner colonizations of the twentieth century evidences how the regulation of the growth of plants involved monitoring and even policing the working population. When growth emanates from the circulation of images into maps, maps into surfaces, and surfaces into images again, power, too, spreads through these circulations.[80] In this regard, the management of the growth of plants not only provided a model for the metabolism of the planet as a whole but it mirrored the escalation of the centralized production of fertilizers and, later, defoliators, which replaced agricultural and vegetal mantles, now synthesized as chemical surfaces.

Aerial images have been described as having "draped the planet with a militarised image of itself."[81] Perhaps the new vegetal mantles that have surfaced can be seen as a planetary cloth.[82] Many other military fabrics would be closely connected, chemically and metaphorically. For example, a photograph documented by Malaysian artist Simryn Gill as part of her research on oil palm plantations presents a series of rows of rubber trees with the caption: "They've tamed our trees and made them march in rows."[83] Landscapes themselves stand as evidence of this militarized image in different ways: either marching as soldiers or hiding as environmental intelligence agents, as in the APT program of the US military; either grown as canopies seeded by bombs launched from the air after the idea of an ex-RAF pilot[84] or as plant formations scrutinized for bioforensic signs of genocide and the aftermath as mass graves.[85]

All such instances are a variation of what we in this book consider the image, or more accurately, "image matter in waiting," in Susan Schuppli's words.[86] Landscapes present themselves as arboreal archives whose photosensitive capacities are constantly actualized by the registering apparatuses they are enmeshed with. While in the words of artist Wiestke Maas, "we can think of photosynthesis itself as a sort of primeval organ of vision spread out across the skin of the vegetable kingdom," a multitude of media and cultural techniques come to the fore framing more specifically such ecological aesthetic.[87] That is the case with transparent glass, measuring light, time-lapses, prescription maps, or machine-learning models, among other examples discussed in the book that have for more than two centuries been part of observing, controlling, and managing how sensing becomes matter. Not by chance, as images have become increasingly active and lifelike entities, the diversity of vegetal matter has receded from the surfaces of the earth. Media are not only concerned with logistical redistribution of the matter taken either as resource or waste but they are also active agents in the transformation of the metabolism of the planet itself. Beyond finitude driven by fossil fuels, media need to be understood in relation to the diversity of surfaces on the planet, as this book has sought to argue.[88] In this task, we concur with Irmgard Emmelhainz as she writes, "The Anthropocene has meant not a new image of the world, but rather a radical change in the conditions of visuality and the subsequent transformation of the world into images."[89] We suggest also reading this point as a modification and deepening of the argument by Walter Benjamin that we referred to in the introduction: "To each truly new configuration of nature—and, at bottom, technology is just such a configuration—there correspond new 'images.'"[90] Such images, thus, are not merely images of the Anthropocene but expressions of their conditions of existence.

Our focus on ecological aesthetics opens up to the uncanny irruption of scalar difference. We have, throughout this book, looked at different cultural techniques involving technical images, but we are as interested in the broader ethical and aesthetic propensities of the forces at play. We are interested in how images can ground new natural contracts through medium design, where long-term assemblages of human communities, environments, and other species participate together, as Gabrys and Tavares have proposed. Cultural techniques can also turn from analysis to speculation. Paraphrasing Karen Barad, the world and its potential for becoming are recursively remade in each cultural technique.[91] Thus, in the "forest of matter"—using Fuller and Goriunova's phrase—different arrangements come to be redesigned and reconceptualized: the glass in the forest, the biosphere in the sensor, the chloroplast in the image, the Lichtgenuss in the data.[92] We always start within an existing assemblage of forces, whether of the forest, prairie, or soil.

This book can be read as a history of environmental mediation and the speculative potential that emerges from the study of specific cases of terrestrial surfaces observed from the point of view of their entanglement with the surface of visual media. What is at stake is a particular view of knowing the earth's surface in order to change it. As we have argued throughout the pages of this book, this project of imaging and transformation has operated on so many scales; it has touched everything from the colonial and imperial, military and geoengineering discourses, while also disentangling the much more subtle perspectives on the progressive potentials. Knowing and changing the surface, we have argued, is not only about plants, territories, and atmospheres, but it is also very much about knowing images and environments of media. As such, the surface inscriptions taken as traces of environmental change are also the starting points for a particular project of recursive planetarity, where ecological aesthetics is the necessary companion to an ecological politics.

Consider, again, the trees on barges floating across the liquid surface that opened the book. As with the glass cases englobing vegetal life, the striking eloquence of those images of vegetal isolation highlights the strength of the invisible ties of the individual plants with their environment. The light through the glass, the barge to the boat: inadvertent logistical chains operate in the background, as part of the volumes of air and water that surround the plants. No storms in sight, no broken glass: a different order stabilizes the continuity between fragility and control, between uncertainty and (weather) prediction. In what sense is a floating tree a surface? It is not only about soil, microbial life, and plant respiration, but also about methods of framing plant life as part of wider envelopes that recursively interface with other subsurface realms of movement and life. Liquid surfaces are often imagined as image surfaces, and in this case, they are also supporting parts of the logistics of floating, moving, transported vegetal life. This surface and the contrasts it brings about are fully enveloped in questions of life, light, and operationalized aesthetics of ecology.

ACKNOWLEDGMENTS

There are several people and institutions we would like to thank for helping to make this research materialize and to turn the research into a book. Our joint thanks go to the colleagues at FAMU at the Academy of Performing Arts. We would like to mention especially our research colleagues in the project Operational Images and Visual Culture for all sorts of feedback and peer support: Tomáš Dvořák, Tereza Stejskalová, Martin Charvát, Michal Šimůnek, Josef Ledvina, Veronika Jirsová, and Silvie Demartini. We also thank Winchester School of Art for various ways our research and practice have been part of that community over the years. Jussi also wants to thank Aarhus University colleagues and the department of Digital Design and Information Studies for the support, including the AUFF project Design and Aesthetics for Environmental Data and the Digital Aesthetics Research Centre (DARC).

Some of the ideas were discussed in several seminars over the years: at the Winchester School of Art, Goldsmiths University of London, Concordia University in Montreal, IKKM in Weimar, UNED in Madrid, Strelka institute's Terraforming program, NTNU (Trondheim), Stockholm University, and UniArts in Helsinki. We are grateful for the very inspiring thinking and practice of the participants there. Also, many ideas were tested in talks by us. For invitations to give such talks, we would like to thank the Instituto Cervantes, Museo Nacional Centro de Arte Reina Sofía, Aarhus Architecture School, University of Southern Denmark (Odense), Bath Spa University, University of Bergen, Linköping University, Edinburgh University, Invisible Landscapes conference, and the Re:Source—Media Art Histories conference 2023 in Venice. We also presented some of the material in conferences like 4S and SLSA; thank you to organizers and audiences.

Different parts of the research have been produced either as residencies or commissions in venues such as Matadero Madrid, Medialab-Prado,

Fundación Cerezales Antonino y Cinia, MUSAC Museum of Contemporary Art of Castilla y Leon, transmediale and Fotomuseum Winterthur.

A list of people to thank includes:

María Andueza, Ryan Bishop, Salomé Jashi, Jane Birkin, Alejandro Limpo, Eda Sancakdar Onikinci, Erkki Huhtamo, Benjamin Bratton, Samir Bhowmik, Nicolay Boyadjiev, Sasha Anikina, Miha Brebenel, Giuliana Bruno, Jara Rocha, Alessandro Ludovico, Asia Bazdyrieva, Solveig Suess, Alfredo Puente, Matthew Fuller, Pasi Väliaho, Rebecca Birch, José María Alagón Laste, Robert Pietrusko, Marcos García, Athina Stamatopoulou, Beny Wagner, Lisa le Feuvre (at the The Holt and Smithson Foundation), Anastasia Kubrak, Christian Ulrik Andersen, Søren Pold, Adnan Hadzi, Magda Tyżlik-Carver, Bruno Marcos, Matiss Groskaufmanis, Antonio Bernacchi, Alicia Lazzaroni, Paolo Patelli, Juan Guardiola, Inés Plasencia, May Ee Wong, Winnie Soon, Manuel Olveira, Lone Koefoed Hansen, Helena Grande, Jannice Käll, Raul Alaejos, Kristoffer Gansing, Bárbara Fluxá, Linda Hilfling Ritasdatter, Santiago Morilla, Yigit Soncul, José Otero, Sandra Santana, Seth Giddings, Yixuan Cai, Seán Cubitt, Emilio López-Galiacho, Jukka Sihvonen, Fernando Broncano, Alberto Murcia, Karen Redrobe, Jeff Scheible, Henrik Bødker, Anette Vandsø, Geoff Cox, Joasia Krysa, Eleonora Roaro, Daphne Dragona, Daniel Fernández, David Prieto, Andrés Rodríguez Muñoz, Marco Rizzetto, Carmen M. Pellicer Balsalobre, Guillermo Cid, Cristobal Gómez Benito, Juan Manuel García Bartolomé, and Rocío Sánchez Serrano from the central archive of the Biblioteca del Ministerio de Agricultura, Pesca y Alimentación, Madrid.

Elise Misao Hunchuck provided her excellent expertise for copyediting this book and helping us to find the right expressions for many of the ideas we hoped to express.

A big thank you also to the staff at the MIT Press for their support and feedback—including the several anonymous peer reviewers whose comments were very helpful. Thank you to the series editor, Seán Cubitt, and the acquisitions editor, Gabriela Bueno Gibbs.

During the research and writing of this book, Timeo and Aimar were born. Abelardo wants to thank his family for their support all this time.

This book has also been supported by Czech Science Foundation funded project 19–26865X "Operational Images and Visual Culture: Media Archaeological Investigations."

An earlier version of chapter 5 was published as:

Abelardo Gil-Fournier and Jussi Parikka, "Ground Truth to Fake Geographies: Machine Vision and Learning in Visual Practices," *AI & Society* 36 (2021): 1253–1262; https://doi.org/10.1007/s00146-020-01062-3 (licensed under a Creative Commons Attribution 4.0 International License).

NOTES

INTRODUCTION

1. Ivan Nechepurenko, "A Love of Trees or a Display of Power? The Odd Park of an Oligarch," *New York Times*, January 17, 2022, https://www.nytimes.com/2022/01/17/world/europe/bidzina-ivanishvili-georgia-trees.html.
2. The Smithson Floating Island was produced by Minetta Brook in collaboration with the Whitney Museum of American Art, the assistance of the Hudson River Park Trust, and the NYC Department of Parks and Recreation in 2005. *Smithson Floating Island*, Balmori, accessed January 15, 2023, http://www.balmori.com/portfolio/smithson-floating-island.
3. On fertilizing Central Park and beyond, see Jane Hutton's *Reciprocal Landscapes: Stories of Material Movements* (Abingdon: Routledge, 2020), 26–62.
4. Robert Smithson, "Frederick Law Olmsted and the Dialectical Landscape," in *Robert Smithson: The Collected Writings*, ed. Jack Flam (Berkeley: University of California Press, 1996), 160. See also Hutton, *Reciprocal Landscapes*, 1–3.
5. We would like to thank Lisa Le Fleuvre from the Holt/Smithson Foundation for contextualizing Smithson's original idea.
6. *An Allegory with a Dog and an Eagle*, ca. 1508–1510, Royal Collection Trust, accessed January 15, 2023, https://www.rct.uk/collection/912496/an-allegory-with-a-dog-and-an-eagle.
7. Aby Warburg, *Mnemosyne Atlas*, panel 48, https://warburg.library.cornell.edu/panel/48.
8. On media archaeological topoi, see Erkki Huhtamo, "Dismantling the Fairy Engine: Media Archaeology as Topos Study," in *Media Archaeology: Approaches, Applications, and Implications*, ed. Erkki Huhtamo and Jussi Parikka (Berkeley: University of California Press, 2011), 27–47.
9. Keller Easterling, *Medium Design. Knowing How to Work on the World* (London: Verso, 2021), 10 and passim.
10. Hutton, *Reciprocal Landscapes*, 3–7.
11. Brett Milligan, "Accelerated and Decelerated Landscapes. On the Techniques, Knowledges, and Ethics of Bending Time," *Places*, February 2022, https://placesjournal.org/article/accelerated-and-decelerated-landscapes/. See also Easterling, *Medium Design*.
12. On a multiscalar approach to the concept of drift, see Bronislaw Szerszynski, "Drift as a Planetary Phenomenon," *Performance Research* 23, no. 7 (2018): 136–144.

13. See, for example, Olga Goriunova and Matthew Fuller, *Bleak Joys: Aesthetics of Ecology and Impossibility* (Minneapolis: University of Minnesota Press, 2019). The scaffolding of our book and its thematic and methodological choices also are based on foundations of artistic practices. Abelardo Gil-Fournier's art has, for a long period, dealt with questions of mediation of environmental surfaces that become interfaces to grasp a plethora of historical, experiential, and material forces of medium design. The visual technical character of some of the agrarian landscapes described in chapter 4 features in Gil-Fournier's installation *Mawat* (2017), and the time-lapse imagery in his *Bildung* (2019) video installation concerning plant growth prompts some of the themes in chapter 6 concerning computation and blades. Similarly, *The Quivering of the Reed* (2019) probes, in installation form, many of the themes of our book: the work puts a living plant into movement and frames it as an *ur*-form of cinema and animation; a sequence of subtitles is projected on the screen describing the images as if they had been filmed on a river, while bringing to mind a quote by Andrei Tarkovsky on his concept of the organic rhythm of the "time-pressure" of cinematic sequences (see Andrey Tarkovsky, *Sculpting in Time: Reflections on the Cinema*, trans. Kitty Hunter-Blair [New York: Alfred A. Knopf, 1987], 113–124). The moving plant becomes the image, the moving image becomes the plant. This is also the theme of our joint work, *Seed, Image, Ground* (2020), a video that is briefly discussed in chapter 4, a take on aerial agricultural practices which tie twentieth-century military views with more recent forms of smart agriculture and data-visualized growth. Of course, while such artistic practices are central to many of the methodological, stylistic, and theoretical views we propose in the book, Gil-Fournier's and our joint work sits as part of a broader burgeoning field of technical imaging, data, and plants. We can here mention also, among other references: the work on agriculture and the politics of space featured in Agnes Denes's *Wheatfield: A Confrontation* (1982) and Regina José Galindo's *Mazorca* (2014); Joan Fontcuberta's vegetal photography of industrial detritus in *Herbarium* (1983), along with Anna Ridler's GAN-driven AI work on tulips and speculative value, *Mosaic Virus* (2019); the aesthetics of measurement featuring in Hans Haacke's *Grass Grows* (1969) and in Marine Hugonnier's video installation *Apicula Enigma* (2013); the enumerated natures in Zygmunt Rytka's video performance series *Continual Infinity* (1982–1993), as well as in Julian Oliver's *psWorld* (2010); the critical approach to the concept of green mantle in Teresa Murak's *Easter Carpet* (1974) and in Julius von Bismark's *Landscape Painting* series (2015–2021); the work on ecosystem services in Helen Mayer Harrison and Newton Harrison's *Hog Pasture: Survival Piece #1* (1970–1971) and in Disnovation group's *Life Support System* (2020); the speculative proposals of Superstudio's *Supersurface: An Alternative Model for Life on the Earth* (1972) and Superflux's *Mitigation of Shock* (2019). In recent design and architectural (speculative) practice, see also Frédérique Aït-Touati, Alexandra Arènes, and Axelle Grégoire, *Terra Forma. A Book of Speculative Maps* (Cambridge, MA: MIT Press, 2022).

14. Among the many fabulous sources on the topic, see, for example, Daniela Bleichmar, *Visible Empire: Botanical Expeditions and Visual Culture in the Hispanic*

Enlightenment (Chicago: University of Chicago Press, 2012); Lucile H. Brockway, *Science and Colonial Expansion: The Role of the British Royal Botanic Gardens* (New Haven, CT: Yale University Press, 2002). Londa Schiebinger and Claudia Swan, eds., *Colonial Botany: Science, Commerce, and Politics in the Early Modern World* (Philadelphia: University of Pennsylvania Press, 2005).

15. Harold Wager, "The Perception of Light in Plants," *Annals of Botany* 23, no. 91 (1909): 459–489. Howard Caygill, "Harold Wager and the Photography of Plants," *Photographies* 14, no. 3 (2021): 505–519.

16. See, for example, Janet Janzen, *Media, Modernity and Dynamic Plants in Early 20th Century German Culture* (Leiden: Brill, 2016); Inga Pollmann, *Cinematic Vitalism: Film Theory and the Question of Life* (Amsterdam: Amsterdam University Press, 2018); as well as the edited collection on this topic, *Puissance du végétal et cinéma animiste: La vitalité révélée par la technique*, ed. Teresa Castro, Perig Pitrou, and Marie Rebecchi (Paris: Les Presses du Réel, 2020).

17. This could be said to be at least partly a cybernetic notion of the planet: not a particular geological form so much as a sphere of interlocking biogeochemical dynamics; ecosystems thinking since the 1930s is one part of the story, another one is how this escalated into systems modeling, both computational and reliant on the variety of sensors that started to make sense of Earth. See Jennifer Gabrys, *Program Earth: Environmental Sensing Technology and the Making of a Computational Planet* (Minneapolis: University of Minnesota Press, 2016). See also Paul N. Edwards, *A Vast Machine: Computer Models, Climate Data, and the Politics of Global Warming*, ed. Geoffrey C. Bowker (Cambridge, MA: MIT Press, 2013). On critical zones, see Bruno Latour and Peter Weibel, eds., *Critical Zones: The Science and Politics of Landing on Earth* (Cambridge, MA: MIT Press, 2020).

18. Sydney Mangham, *Earth's Green Mantle: Plant Science for the General Reader* (London: English Universities Press, 1939), 32. See also Veronica della Dora, *The Mantle of the Earth: Genealogies of a Geographical Metaphor* (Chicago: University of Chicago Press, 2020).

19. William Rankin, *After the Map: Cartography, Navigation, and the Transformation of Territory in the Twentieth Century* (Chicago: University of Chicago Press, 2016), color gallery page 2.

20. Matt Edgeworth, "The Relationship between Archaeological Stratigraphy and Artificial Ground and Its Significance in the Anthropocene," in *A Stratigraphical Basis for the Anthropocene*, ed. Colin Neil Waters, Jan Zalasiewicz, Mark Williams, Michael A. Ellis, and Andrea M. Snelling (London: Geological Society of London, 2014), 91–108, at 105, quoted in Stephen Graham, *Vertical: The City from Satellites to Bunkers* (London: Verso, 2016).

21. On afforestation and (landscape) design, for example, see Rosetta S. Elkin, *Plant Life: The Entangled Politics of Afforestation* (Minneapolis: University of Minnesota Press, 2022).

22. On medianatures, see Jussi Parikka, *A Geology of Media* (Minneapolis: University of Minnesota Press, 2015). See also Bernard Stiegler's *Bifurcate: There Is No*

Alternative, ed. Bernard Stiegler and the International Collective, trans. Daniel Ross (London: Open Humanities Press, 2021).

23. Hutton, *Reciprocal Landscapes*, 11.
24. Rankin, *After the Map*.
25. Here the Landsat satellite system since the 1970s was instrumental in opening up such practices of sensing to broader agricultural and other sectors. See Pamela E. Mack, *Viewing the Earth: The Social Construction of the Landsat Satellite System* (Cambridge, MA: MIT Press, 1990). See also Gökçe Önal, "Media Ecologies of the 'Extractive View' Image Operations of Material Exchange," *Footprint* 14, no. 2 (2020): 31–48.
26. Kathryn Yusoff, "Excess, Catastrophe, and Climate Change," *Environment and Planning D: Society and Space* 27, no. 6 (2009): 1010–1029, at 1010.
27. Geoff Manaugh, "Geomedia, or What Lies Below," *BLDGBLOG*, December 31, 2020. https://bldgblog.com/2020/12/geomedia-or-what-lies-below/. See also Seán Cubitt, "Three Geomedia," *Ctrl-Z* 7 (2017), accessed January 15, 2023, http://www.ctrl-z.net.au/articles/issue-7/cubitt-three-geomedia/. On the "subsurface," see Karen Pinkus, *Subsurface* (Minneapolis: University of Minnesota Press, 2023).
28. Ros Gray and Shela Sheikh, "The Wretched Earth: Botanical Conflicts and Artistic Interventions," *Third Text* 32, no. 2–3 (2018): 163–175, One could even go so far as to read our book as a variation on what Paul Virilio saw as the fundamental perceptual force of military technologies, especially when introduced to "the blinding Hiroshima flash which literally photographed the shadow cast by beings and things, so that every surface immediately became war's recording surface, its film." Paul Virilio, *War and Cinema: The Logistics of Perception*, trans. Patrick Camiller (London: Verso, 1989), 85. See also Akira Mizuta Lippit, *Atomic Light (Shadow Optics)* (Minneapolis: University of Minnesota Press, 2005). On herbicidal warfare, see David Zierler, *The Invention of Ecocide: Agent Orange, Vietnam, and the Scientists Who Changed the Way We Think about the Environment* (Athens: University of Georgia Press, 2011).
29. Siegfried Kracauer, *The Mass Ornament: Weimar Essays*, trans. and ed., with an introduction by Thomas Y. Levin (Cambridge, MA: Harvard University Press, 1995), 75.
30. Clement Greenberg, "On the Role of Nature in Modernist Painting," in *Art and Culture: Critical Essays* (Boston, MA: Beacon, 1961), 172.
31. Greenberg, "On the Role of Nature in Modernist Painting," 173. This was also the military aesthetic of the aerial view pertinent to Cubism: "The impression that the earth has become a planar grid for abstract data plotting is, in turn, visually enhanced by the flatness of the high-altitude vertical view, with its paradoxical erasure of the vertical dimension; and that leveling might further suggest cubist rejections of perspectival convention in favor of a conspicuously two-dimensional image plane whose depthlessness, like the aerial view's, was no longer hospitable to the conventional pictorial distinction between figure and ground." Paul K. Saint-Amour, "Modernist Reconnaissance," *Modernism/Modernity* 10, no. 2 (2003): 352.

32. Giuliana Bruno, *Surface: Matters of Aesthetics, Materiality, and Media* (Chicago: University of Chicago Press, 2014); Esther Leslie, *Synthetic Worlds: Nature, Art and the Chemical Industry* (London: Reaktion, 2005); Celia Lury, Luciana Parisi, and Tiziana Terranova, "Introduction: The Becoming Topological of Culture," in *Theory, Culture & Society* 29, no. 4–5 (2012): 3–35; Janet Ward, *Weimar Surfaces: Urban Visual Culture in 1920s Germany* (Berkeley: University of California Press, 2001); Yeseung Lee, ed., *Surface and Apparition: The Immateriality of Modern Surface* (London: Bloomsbury, 2020); Tim Ingold, "Surface Visions," *Theory, Culture & Society* 34, no. 7–8 (2017): 99–108.

33. Bruno, *Surface*. See also her more recent book in which issues of art and projection are continued: Giuliana Bruno, *Atmospheres of Projection: Environmentality in Art and Screen Media* (Chicago: University of Chicago Press, 2022).

34. Bruno, *Surface*, 112

35. The surface-screen double becomes architectural in media artistic works such as Janet Cardiff and George Bures Miller's sonic environment installation *The Paradise Institute* (2003) and in projects such as Diller Scofidio + Renfro's cloud building (Blur Building, Yverdon-les-Bains, Switzerland) of 2002; Bruno, *Surface*, 134–137. Expanding on the atmospheric, see also the fog sculptures by Nakaya Fujiko for the Pepsi Pavilion at the Expo '70, discussed by Yuriko Furuhata in *Climatic Media: Transpacific Experiments in Atmospheric Control* (Durham, NC: Duke University Press, 2022), 27–31.

36. The focus on atmosphere as fundamental to contemporary politics is a theme that Peter Sloterdijk has developed. See, for example, Peter Sloterdijk, *Terror from the Air*, trans. Amy Patton and Steve Corcoran (Los Angeles, CA: Semiotext(e), 2009).

37. Lukáš Likavčan and Paul Heinicker, "Planetary Diagrams: Towards an Autographic Theory of Climate Emergency," in *Photography Off the Scale: Technologies and Theories of the Mass Image*, ed. Tomáš Dvořák and Jussi Parikka (Edinburgh: Edinburgh University Press, 2021), 211–230. See also Dietmar Offenhuber, "The Planet as a Photographic Plate," *Fotograf Magazine* 20, no. 40 (2021): 66–71. Also, some approaches in geography and architecture, environmental humanities, and the work on the feral ecologies in and out of soil have defined surfaces beyond flatness. See, for instance, John May, "Logic of the Managerial Surface," *PRAXIS* 13 (2012): 116–124; Robert G. Pietrusko, "The Surface of Data," *LA+* 4 (Fall 2016): 79–83; María Puig de la Bellacasa, *Matters of Care: Speculative Ethics in More than Human Worlds* (Minneapolis, MN: University of Minnesota Press, 2017), and Anna L. Tsing, *The Mushroom at the End of the World: On the Possibility of Life in Capitalist Ruins* (Princeton, NJ: Princeton University Press, 2015).

38. On the work of Jananne al-Ani, see Caren Kaplan, *Aerial Aftermaths: Wartime from Above* (Durham, NC: Duke University Press, 2018), 180–206.

39. Eyal Weizman, *Forensic Architecture: Violence at the Threshold of Detectability* (New York: Zone, 2017), 274. See also Susan Schuppli, *Material Witness: Media, Forensics, Evidence* (Cambridge, MA: MIT Press, 2020).

40. Adrian J. Ivakhiv, *Ecologies of the Moving Image: Cinema, Affect, Nature* (Waterloo: Wilfried Laurier University Press, 2013), 6. For a related argument on the zoetic

character of photography see also Joanna Zylinska, *Nonhuman Photography: Theories, Histories, Genres* (Cambridge, MA: MIT Press, 2017).

41. Graig Uhlin, "Plant-Thinking with Film: Reed, Branch, Flower," in *The Green Thread: Dialogues with the Natural World*, ed. Patrícia Vieira, Monica Gagliano, and John Ryan (Lanham, MD: Lexington, 2016), 201–218, at 203.

42. Nadia Bozak, *The Cinematic Footprint: Lights, Camera, Natural Resources* (New Brunswick, NJ: Rutgers University Press, 2012), 13, 18. "Each film frame is a measure of our civilization's control of the sun, in the form of the fossilized sun or carbon that we have captured, refined, and duly exploited. The cinematic image literalizes in incontrovertible terms, how industry, the images of industrial culture, and the earth's natural ecology are, together and on their own terms, categorically derived from the power that emanates from the sun's rays" (29).

43. James Corner, "Terra Fluxus," in *The Landscape Urbanism Reader*, ed. Charles Waldheim (New York: Princeton Architectural Press, 2005), 21–33. See also Brett Milligan, "Landscape Migration. Environmental Design in the Anthropocene," *Places Journal*, June 2015. https://placesjournal.org/article/landscape-migration/#0, and Alfred W. Crosby, *Ecological Imperialism: The Biological Expansion of Europe, 900–1900* (Cambridge: Cambridge University Press, 2004), on the colonial amplification of such drifts.

44. See for instance the case of the moving borders in the Alps between Italy and its neighbor countries in Marco Ferrari, Elisa Pasqual, and Andrea Bagnato, eds., *A Moving Border: Alpine Cartographies of Climate Change* (New York: Columbia University Press, 2018).

45. See Abelardo Gil-Fournier and Jussi Parikka. "'Visual Hallucination of Probable Events' On Environments of Images, Data, and Machine Learning," in *Big Data: A New Medium?*, ed. Natasha Lushetich (Abingdon: Routledge, 2021), 46–60.

46. Walter Benjamin, *The Arcades Project*, trans. Howard Eiland and Kevin McLaughlin (Cambridge, MA: Belknap Press of Harvard University Press, 1996), 390 (Kla, 3).

47. Anusuya Datta, "Machine Learning Creates Living Atlas of the Planet," *Geospatial World* (blog), April 11, 2017. https://www.geospatialworld.net/article/machine-learning-creates-living-atlas-of-the-planet/.

48. See Jussi Parikka, *A Geology of Media* (Minneapolis: University of Minnesota Press, 2015). See also Jussi Parikka, "Medianatures," in *Posthuman Glossary*, ed. Rosi Braidotti and Maria Hlavajova (London: Bloomsbury Academic, 2018), 251–253. Related accounts appear also for example in Ivakhiv, *Ecologies of the Moving Image*, as well as in Adam Wickberg and Johan Gärdebo, eds., *Environing Media* (Abingdon: Routledge, 2023). Also compare Seán Cubitt's words on mediation as a primal flow: "It is the task of media theory to distinguish the philosophical concept of mediation as primal flow connecting all things from the history of mediation as the engine of disconnection and delay, and thus of exploitation and oppression." Seán Cubitt, *The Practice of Light: A Genealogy of Visual Technologies from Prints to Pixels* (Cambridge, MA: MIT Press, 2014), 2–3.

49. "So-called nature" is a riff on Friedrich Kittler's "so-called Man," which is a figure underpinned by the technological conditioning of such notions in the humanities, and which thus also historicizes such figures as part of cultural techniques that produce them. In Kittler's case that technological conditioning would include psychophysics, writing technologies, audio, cinema, and so on. There is more than a hint of Foucault at the back of Kittler's concept. See Friedrich A. Kittler, *Discourse Networks, 1800/1900*, trans. Michael Metteer with Chris Cullens (Stanford, CA: Stanford University Press 1990).

50. See Cubitt, *Three Geomedia*. This merger of the material and epistemic sides of mediation speaks to the new materialist agenda put forward by, for example, Karen Barad. It also links to John Durham Peters' work where elemental media is discussed not only regarding the environmental impact of media technologies and infrastructures but even more by going back to the large-scale mediations from sun to the sky and waters to nonhuman temporalities that determine a broader way of appreciating some "deep times" of media. John Durham Peters, *The Marvelous Clouds: Toward a Philosophy of Elemental Media* (Chicago: University of Chicago Press, 2015); Karen Barad, *Meeting the Universe Halfway: Quantum Physics and the Entanglement of Matter and Meaning* (Durham, NC: Duke University Press, 2007). See also Richard Grusin, "Radical Mediation," *Critical Inquiry* 42, no. 1 (2015): 124–148.

51. On cultural techniques, see, for example, Geoffrey Winthrop-Young, "Cultural Techniques: Preliminary Remarks," *Theory, Culture & Society* 30, no. 6 (2013): 3–19. Thomas Macho, "Second-Order Animals: Cultural Techniques of Identity and Identification," *Theory, Culture & Society* 30, no. 6 (2013): 30–47.

52. Bernhard Siegert, *Cultural Techniques: Grids, Filters, Doors, and Other Articulations of the Real*, trans. Geoffrey Winthrop-Young (New York: Fordham University Press, 2015), 14

53. Cornelia Vismann, "Cultural Techniques and Sovereignty," *Theory, Culture & Society* 30, no. 6 (2013): 83–93, at 84. Every plan, infrastructure, and media operation comes to be defined, even after the fact, by its own execution. Hence the idea voiced by Vismann, that cultural techniques deal with the middle-voice and the verb form of action, is central to this focus, where "all media and things supply their own rules of execution." All contributing agents, participants, and objects, including those of knowledge, emerge from the primacy of productive cultural techniques. "To derive the operational script from the resulting operation, to extract the rules of execution from the executed act itself: that is what characterizes the approach of cultural techniques"; Vismann, "Cultural Techniques and Sovereignty," 87.

54. Siegert, *Cultural Techniques*, 149.

55. On the invisual, see Adrian MacKenzie and Anna Munster, "Platform Seeing: Image Ensembles and Their Invisualities," *Theory, Culture & Society* 36, no. 5 (2019): 3–22. See also Jussi Parikka, *Operational Images: From the Visual to the Invisual* (Minneapolis: University of Minnesota Press, 2023). Due to our focus in this book, we leave aside experiments related to genes, hormones, or electric

signals. During the same period, when some of the experiments discussed in the book took place, other investigations analyzed plants from the point of view of genetics, hormone regulation, and electric signals. On the link between Darwin's research into plant movement as the basis for plant hormones and the later work on the chemistry of plant hormones, see David Zierler, *The Invention of Ecocide: Agent Orange, Vietnam, and the Scientists Who Changed the Way We Think about the Environment* (Athens: University of Georgia Press, 2011). On experimental cultures related to breeding and genetic research, see Christophe Bonneuil, "Mendelism, Plant Breeding and Experimental Cultures: Agriculture and the Development of Genetics in France," *Journal of the History of Biology* 39, no. 2 (2006): 281–308. On interaction with plants through electric signals, see Stefan Rieger, "What's Talking? On the Nostalgic Epistemology of Plant Communication," in Ryan, Vieira, and Gagliano, *Green Thread*, 59–79.

56. See, for instance, Sebastian Vehlken, *Zootechnologies: A Media History of Swarm Research* (Amsterdam: Amsterdam University Press, 2019). Furthermore, the question of cultural techniques has an interesting relation to themes of urbanism and the rural, as it underpins the approach as to where "media" or "culture" is found. On related discussions, see Shannon Mattern on the *longue durée* of urbanization in *Code and Clay, Data and Dirt: Five Thousand Years of Urban Media* (Minneapolis: University of Minnesota Press, 2017). See also Manuel DeLanda, *A Thousand Years of Nonlinear History* (New York: Zone, 2000). On ruralism, see Neil Brenner and Nikos Katsikis, "Hinterlands of the Capitalocene," *Architectural Design* 90 (2020): 22–31.

57. Siegert, *Cultural Techniques*, 145.

58. See Jara Rocha and Femke Snelting, "So-called Plants," in *Volumetric Regimes: Material Cultures of Quantified Presence*, ed. Jara Rocha and Femke Snelting (London: Open Humanities Press, 2022), 239–260. In a similar way, cultural techniques research has related more closely to anthropotechnics that have given rise to humans-as-cultural-beings and to domesticated animals. For a discussion on cultural techniques in relation to anthropotechnics and other posthuman anthropological considerations, see Winthrop-Young, "Cultural Techniques." Cultural techniques produce the subjects and objects they operate with. In this regard, it is a media theoretical approach related to Karen Barad's onto-epistemology, where "knowing is a direct material engagement, a practice of intra-acting with the world as part of the world in its dynamic material configuring, its ongoing articulation. The entangled practices of knowing and being are material practices." See Barad, *Meeting the Universe Halfway*, 379. For a more metaphysical approach in plant studies, see Emanuele Coccia, *The Life of Plants: A Metaphysics of Mixture* (Medford, MA: Polity, 2018). Compare also Wietske Maas's proposition that "we can think of photosynthesis itself as a sort of primeval organ of vision spread out across the skin of the vegetable kingdom"; Wietske Maas, "The Corruption of the Eye: On Photogenesis and Self-Growing Images," *e-flux*, 56, Venice Biennale (September 2015), accessed January 15, 2023, http://supercommunity.e-flux.com/texts/the-corruption-of-the-eye-on-photogenesis-and-self-growing-images/.

59. See Michael Marder, *Plant-Thinking: A Philosophy of Vegetal Life* (New York: Columbia University Press, 2013), and Monica Gagliano, John C. Ryan, and Patrícia Vieira, eds., *The Language of Plants: Science, Philosophy, Literature* (Minneapolis: University of Minnesota Press, 2017). For texts coming from the biological sciences, see, for instance, Anthony Trewavas, *Plant Behaviour and Intelligence* (Oxford: Oxford University Press, 2015), and Stefano Mancuso and Alessandra Viola, *Brilliant Green: The Surprising History and Science of Plant Intelligence*, trans. Joan Benham (Washington, DC: Island Press, 2015); and Jeffrey Nealon, *Plant Theory: Biopower and Vegetable Life* (Stanford, CA: Stanford University Press, 2015).

60. See Giovanni Aloi, ed., *Botanical Speculations: Plants in Contemporary Art* (Newcastle upon Tyne: Cambridge Scholars, 2018). On bioart, see George Gessert, *Green Light: Toward an Art of Evolution*, (Cambridge, MA: MIT Press, 2010). For an approach to the topic of plants as queer entities in relation to art and film, see Teresa Castro, "The Mediated Plant," *e-flux* 102 (September 2019), accessed January 15, 2023, https://www.e-flux.com/journal/102/283819/the-mediated-plant/.

61. See Janzen, *Media, Modernity and Dynamic Plants*; Uhlin, "Plant-Thinking with Film"; Matthew Vollgraff, "Vegetal Gestures: Cinema and the Knowledge of Life in Weimar Germany," *Grey Room* 72 (Summer 2018): 68–93; Oliver Gaycken, "The Secret Life of Plants: Visualizing Vegetative Movement, 1880–1903," in *Early Popular Visual Culture* 10, no. 1 (February 1, 2012): 51–69. See also Katharina Steidl, "Leaf Prints. Early Cameraless Photography and Botany," *PhotoResearcher* 17 (2012): 26–35, and Max Long, "The Ciné-Biologists: Natural History Film and the Co-Production of Knowledge in Interwar Britain," *British Journal for the History of Science* 53, no. 4 (2020): 527–551.

62. Pollmann, *Cinematic Vitalism*; Castro, Pitrou, and Rebecchi, *Puissance du végétal et cinéma animiste*; Hannah Landecker, "Microcinematography and the History of Science and Film," *Isis* 97, no. 1 (2006): 121–132. On nonhuman photography, see Zylinska, *Nonhuman Photography*; see also Claire Colebrook, *Deleuze and the Meaning of Life* (London: Continuum, 2010). This is also the key focus of artist and scholar Beny Wagner's work, including his PhD research on metabolism and cinema: Wagner, "Metabolisms of the Moving Image," PhD diss., University of Southampton, September 2023. See also the film by Sasha Litvintseva and Beny Wagner, *My Want of You Partakes of Me* (2023).

63. Susan Schuppli refers here to some of the epistemic uses of the idea of natural sensors, of environmental monitoring through leaves, plants, and vegetal formations. Schuppli, *Material Witness*, 287.

64. Note that such views of plants "conceived of as emitters and receivers of information, following the computational standard of intelligence," are often seen as reductive in the ecocritical articulations of plant studies. See Michael Marder, "To Hear Plants Speak," in Gagliano, Ryan, and Vieira, *Language of Plants*, 103–125, at 115.

65. Consider, for example, the Normalized Difference Vegetation Index (NDVI) as one such operative technique. NDVI is an inconspicuous device for quantification and "coloring," interpreting satellite or other aerial measurements of visible

and invisible light to produce color charts that indicate important information about landcover, such as soil, water, vegetal canopy areas, and importantly, health of vegetation. While the use of infrared film emerged in an earlier period of aerial imaging and had military uses (to detect camouflage), it was early on understood to be useful also for reading the health status of vegetation (Mack, *Viewing the Earth*, 32–33). In subsequent earth observation satellite measurement practices, such as Landsat satellites, the NDVI emerged as one crucial technique of quantification that is often defined simply as quantification of "vegetation by measuring the difference between near-infrared (which vegetation strongly reflects) and red light (which vegetation absorbs)"; GISGeography, 2022, https://gisgeography.com/ndvi-normalized-difference-vegetation-index/. Techniques such as NDVI standardize how color is categorized in the sense data of the earth surface concerning surface biome signatures. For water bodies, different algorithmic devices are used, such as Floating Algae Index (FAI), to help detect algae and to create relevant false-color maps for operational uses. See Melody Jue, "'Pixels May Lose Kelp Canopy': The Photomosaic as Epistemic Figure for the Satellite Mapping and Modeling of Seaweeds," *Media+Environment* 3, no. 2 (2021), https://mediaenviron.org/article/21261-pixels-may-lose-kelp-canopy-the-photomosaic-as-epistemic-figure-for-the-satellite-mapping-and-modeling-of-seaweeds. The quantified coloring of the surface adds a layer of data that is premised on the sensing of and by plants. Such knowledge objects as NDVI emerge as central "administrative" elements in the mediation of light from photosynthesis to quantification of remote sensing data. It also becomes the backbone of scientific observation and environmental services as well as precision agriculture as it is practiced as a data-driven high-tech business. Enumeration and statistical knowledge of surfaces become inextricably linked to fluctuating shades of green; what arose initially as the plant's own primary sensing and absorption of light, the color has become a technical element of the analysis.

66. Stefan Ouma, *Farming as Financial Asset: Global Finance and the Making of Institutional Landscapes* (Newcastle: Agenda, 2020). See also Xiaowei Wang, *Blockchain Chicken Farm: And Other Stories of Tech in China's Countryside* (New York: Macmillan, 2020); Adam Wickberg, "Environing Media and Cultural Techniques: From the History of Agriculture to AI-Driven Smart Farming," *International Journal of Cultural Studies* 26, no. 4 (2023): 392–409, https://doi.org/10.1177/13678779221144762. On plants as dead labor, see Bozak, *Cinematic Footprint*; Luce Lebart, "La photographie d'origine végétale," in Castro, Pitrou, and Rebecchi, *Puissance du végétal et cinéma animiste*, 119–127. On solar energy and bioinfrastructures, see also Puig de la Bellacasa, *Matters of Care*; Nicole Starosielski. "Beyond the Sun: Embedded Solarities and Agricultural Practice," *South Atlantic Quarterly* 120, no. 1 (2021): 13–24.

67. The managerial surface is defined as the technical media control space that governs the new territorial arrangements in electro-statistical media—including digital modes of visualization, tracking, simulation, modelling. See May, "Logic of the Managerial Surface."

68. Already for the twentieth-century imaginary of environmental management, this meant developing systems of sensing that were able to interpret such signs for the development of prediction models, from sensors onboard Landsat satellite in 1970s to more recent precision farming data management plans. See Mack, *Viewing the Earth*.

69. See, for instance, Suzanne Simard, *Finding the Mother Tree: Discovering the Wisdom of the Forest* (New York: Alfred A. Knopf, 2021), and Anna-Sophie Springer, Etienne Turpin, Kirsten Einfeldt, and Daniela Wolf, eds., *The Word for World Is Still Forest* (Berlin: HKW Verlag, 2017). For a biosemiotic take on the environment built by a forest in relation also to its human dwellers, see also Eduardo Kohn, *How Forests Think: Toward an Anthropology beyond the Human* (Berkeley: University of California, 2013).

70. See in this regard Elaine P. Miller, *The Vegetative Soul: From Philosophy of Nature to Subjectivity in the Feminine* (Albany: SUNY Press, 2002), and Michael Pollan, *The Botany of Desire: A Plant's-Eye View of the World* (New York, NY: Random House, 2002). See also Tsing, *Mushroom at the End of the World*.

71. Jennifer Gabrys, "Smart Forests and Data Practices: From the Internet of Trees to Planetary Governance," *Big Data & Society* 7, no. 1 (2020). See also Max Liboiron, *Pollution Is Colonialism* (Durham, NC: Duke University Press, 2021), for a related merger of knowledge practices in relation to the study of plastic pollution in water.

72. See for instance Paulo Tavares, "The Geological Imperative: On the Political Ecology of Amazonia's Deep History," in *Architecture in the Anthropocene: Encounters among Design, Deep Time, Science and Philosophy*, ed. Etienne Turpin (London: Open Humanities Press, 2013), 209–240, and Paulo Tavares, "Nonhuman Rights," in *Forensis: The Architecture of Public Truth*, ed. Forensic Architecture (Berlin: Sternberg, 2014), 553–572.

73. Jennifer Gabrys. "Becoming Planetary," *e-flux*, October 2018, https://www.e-flux.com/architecture/accumulation/217051/becoming-planetary/.

74. Gabrielle Hecht, "Interscalar Vehicles for an African Anthropocene: On Waste, Temporality, and Violence," *Cultural Anthropology* 33, no. 1 (2018): 109–141.

75. Kliment A. Timiriazev, "Croonian Lecture. The Cosmical Function of the Green Plant," *Proceedings of the Royal Society of London* 72, nos. 477–486 (1904): 424–461.

76. On the energy and geopolitics of the Soviet Russia as backdrops to questions of "planetarity," see Daniela Russ, "'Socialism Is Not Just Built for a Hundred Years': Renewable Energy and Planetary Thought in the Early Soviet Union (1917–1945)," *Contemporary European History* 31 (2022): 491–508.

77. See also Matthew Fuller and Eyal Weizman, *Investigative Aesthetics: Conflicts and Commons in the Politics of Truth* (London: Verso, 2021), and Schuppli, *Material Witness*.

CHAPTER 1

1. Giacomo Ciamician, "The Photochemistry of the Future," *Science* 36, no. 926 (1912): 385–394.
2. Ciamician, "Photochemistry of the Future," 394.
3. Ciamician, "Photochemistry of the Future," 394.
4. Achille Mbembe discusses the slavery system as one of energy, of ambulant suns and fossils. "The ecological disturbances brought about by this vast draining of humans and its procession of violence have yet to be systematically studied. But the New World plantations could hardly have operated without the massive use of 'ambulant suns,' that is, African slaves. Even after the Industrial Revolution, these real human fossils continued to serve as coal for producing energy and provided the necessary dynamism for economically transforming the Earth System." Mbembe, *Necropolitics*, trans. Steven Corcoran (Durham, NC: Duke University Press, 2019), 165–166.
5. Marianne Klemun, "Live Plants on the Way: Ship, Island, Botanical Garden, Paradise and Container as Systemic Flexible Connected Spaces in between," *HOST—Journal of History of Science and Technology* 5 (Spring 2012), http://johost.eu/vol5_spring_2012/marianne_klemun_2.htm. Londa Schiebinger, *Plants and Empire: Colonial Bioprospecting in the Atlantic World* (Cambridge, MA: Harvard University Press, 2007).
6. Ciamician, "Photochemistry of the Future," 394.
7. To consider the classic example, namely Paul Scheerbart's *Glasarchitektur* [Glass architecture] (Berlin: Verlag der Sturm, 1914), which painted a picture of transformations of not only the architectural materials from concrete to the iron and glass combinations of urban and garden buildings but also the subsequent speculative terraforming of the surface of the earth itself: "We all know what is meant by colour; it forms only a small part of the spectrum. But we want to have that part. Infra-red and ultra-violet are not perceptible to our eyes—but ultra-violet is perceptible to the sensory organs of ants. If we cannot at the moment accept that our sensory organs will develop appropriately overnight, we are justified in accepting that we should first reach for what is within our grasp—i.e., that part of the spectrum which we are able to in with our own eyes—in fact, the miracles of colour, which we are in a position to appreciate ourselves. In this, only glass architecture, which will inevitably transform our whole lives and the environment in which we live, is going to help us. So we must hope that glass architecture will indeed transform the face of our world." Paul Scheerbart in *Glass Architecture by Paul Scheerbart and Alpine Architecture by Bruno Taut*, trans. Shirley Palmer and James C. Palmes (Westport, CT: Praeger, 1972), 70.
8. Gabrielle Hecht, "Interscalar Vehicles for an African Anthropocene: On Waste, Temporality, and Violence," *Cultural Anthropology* 33, no. 1 (2018): 109–141.
9. Hecht, "Interscalar Vehicles for an African Anthropocene," 115.

10. Seán Cubitt, *The Practice of Light. A Genealogy of Visual Technologies from Prints to Pixels* (Cambridge, MA: MIT Press 2014).
11. On these practices of "environing" and their relation to the concept of environment itself at the planetary scale, see Sabine Höhler, "Ecospheres: Model and Laboratory for Earth's Environment," *Technosphere*, June 20, 2018. https://technosphere-magazine.hkw.de/p/Ecospheres-Model-and-Laboratory-for-Earths-Environment-qfrCXdpGUyenDt224wXyjV.
12. Friedrich A. Kittler, "Computer Graphics: A Semi-Technical Introduction," trans. Sara Ogger, *Grey Room* 2 (Winter 2001): 30–45, at 38.
13. See Jacob Gaboury, *Image Objects: An Archaeology of Computer Graphics* (Cambridge, MA: MIT Press, 2021), 27–54.
14. Light was for Descartes invisible, insubstantial, and static. His faith in the ontological truth of premises such as the axioms of Euclidean geometry, however, made it possible for the philosopher to represent the paths of light through metaphors such as line drawings—"the line is a tool for the representation of light and a description of its behavior", argues Cubitt—making description and representation coalesce. See Cubitt, *Practice of Light*, 56.
15. Cubitt, *Practice of Light*, 58.
16. Cubitt, *Practice of Light*, 77.
17. Edward Grant, *Much Ado about Nothing: Theories of Space and Vacuum from the Middle Ages to the Scientific Revolution* (Cambridge: Cambridge University Press, 2011), 67–100.
18. "The centre of interest in these questions in 1645–1651 was France, where Mersenne reported on the Italian work, and where natural philosophers such as Pascal, Petit, Roberval, and Pecquet all gave their views and experimented with the Torricellian apparatus." Steven Shapin and Simon Schaffer, *Leviathan and the Air-Pump* (Princeton, NJ: Princeton University Press, 1985), 41.
19. As the well-known book by the historians of science Steven Shapin and Simon Schaffer has shown in detail, the inverted glass jar with candles inside or connected to an air-pump succeeded in the seventeenth century in becoming the model of legitimate production of science—based on the value of systematic experimental practices—after a public struggle between experimentalist Robert Boyle and natural philosopher Thomas Hobbes. See Shapin and Schaffer, *Leviathan and the Air-Pump.*
20. See Laura Baudot, "An Air of History: Joseph Wright's and Robert Boyle's Air Pump Narratives," *Eighteenth-Century Studies* 46, no. 1 (2012): 1–28.
21. Joseph Priestley, *Experiments and Observations on Different Kinds of Air* (London: J. Johnson, 1775), 50. "Noxious air" was carbon dioxide, as it was addressed by Priestley. See also 89–90.
22. Priestley, *Experiments and Observations*, 50. See also Joe Jackson, *A World on Fire: A Heretic, an Aristocrat, and the Race to Discover Oxygen* (New York: Viking Penguin, 2005), 106–107.

23. Priestley, *Experiments and Observations*, 86

24. The president of the Royal Society, Sir John Pringle, on the occasion of presentation of the Copley medal to Priestley. Kliment A. Timiriazev, *The Life of the Plant: Ten Popular Lectures* (Moscow: Foreign Languages Publishing House, 1958), 395.

25. Priestley, *Experiments and Observations*, 93.

26. Kijan Malte Espahangizi, "Wissenschaft im Glas: Eine historische Ökologie moderner Laborforschung," PhD diss., ETH Zürich, 2010, 38.

27. This exchange is related to an observation that John D. Peters makes in relation to an experiment by James C. Maxwell, where two lenses in contact are shown to be infinitesimally distanced as they produce a diffraction pattern that only distance would be able to generate. This episode is analyzed by Peters in relation to the modern Western subject that at the end of the nineteenth century faced a wide spectrum of media for communication while at the same time felt deeply the experience of isolation. It highlights the tension between the inevitable presence of vacuum between bodies and objects and the resort to communication to mitigate the anxiety of solipsism. In this regard, in Priestley's case light operates as an elemental form of communication where the communicated message is the continuity of the inhabitable world. See John D. Peters, *Speaking into the Air: A History of the Idea of Communication* (Chicago: University of Chicago Press, 2001). A discussion on Peters observation can be found also in Sybille Krämer, *Medium, Messenger, Transmission: An Approach to Media Philosophy* (Amsterdam: Amsterdam University Press, 2015).

28. Schiebinger, *Plants and Empire*, 89.

29. Luke Keogh, *The Wardian Case: How a Simple Box Moved Plants and Changed the World* (Chicago: University of Chicago Press, 2020), 32

30. This is well reflected in such works as John Woodward's *Brief Instructions for Making Observations in All Parts of the World; as also for Collecting, Preserving and Sending Over Natural Things* from 1696. See also Keogh, *Wardian Case,*, 33.

31. Zeynep Çelik Alexander, "Managing Iteration: The Modularity of the Kew Herbarium," in *Iteration: Episodes in the Mediation of Art and Architecture*, ed. Robin Schuldenfrei (Abingdon: Routledge, 2020), 1–24.

32. Nathaniel B. Ward, *On the Growth of Plants in Closely Glazed Cases*, 2nd ed. (London: John van Voorst, Paternoster Row, 1852), 72. The first edition came out in 1842.

33. The case can be seen as one emblem of the broader inspiration provided by glass materials in the nineteenth century. See Isobel Armstrong, *Victorian Glassworlds: Glass Culture and the Imagination, 1830–1880* (Oxford: Oxford University Press, 2008). As Margaret Flanders Darby notes on her take on the "unnatural history" of the period: "As parlor accessory, Ward's invention is both an extreme and characteristic the Victorians' artificial manipulation of nature: a portable, frugal expression collecting manias, of the development of the private sphere, of a response pollution that depended on the very innovative industrial

technologies that problem." See Margaret Flanders Darby, "Ward's Glass Cases," *Victorian Literature and Culture* 35, no. 2 (2007): 635–647, at 647. The spread of these terraria and aquaria in private spheres continued beyond the Victorian context; on aquaria, in particular, see Christina Wessely and Thomas Brandstetter, eds., "Der Ozean im Glas: Aquaristische Räume um 1900," special issue, *Berichte zur Wissenschaftsgeschichte* 2, no. 36 (2013).

34. Wolfgang Schivelbusch, *The Railway Journey: The Industrialization of Time and Space in the Nineteenth Century* (Berkeley: University of California Press, 2014), 78.

35. Lucae, quoted in Schivelbusch, *Railway Journey*, 79.

36. Ward, *On the Growth of Plants*. On plant indicators, see chapter 6.

37. Ward, *On the Growth of Plants*, 38.

38. Faraday letter to Ward, November 4, 1851, included in the appendix of Ward, *On the Growth of Plants*.

39. See Alexander, "Managing Iteration." On the concept of center of calculation, see Bruno Latour, *Science in Action: How to Follow Scientists and Engineers through Society* (Cambridge, MA: Harvard University Press, 1988).

40. See Michel Foucault, "Of Other Spaces." Trans. Jay Miskowiec, *Diacritics* 16, no. 1 (1986): 22–27. "The heterotopia is capable of juxtaposing in a single real place several spaces, several sites that are in themselves incompatible" (25).

41. Keogh, *Wardian Case*, 181.

42. Michel Serres, *The Parasite*, trans. Lawrence R. Schehr (Baltimore, MD: Johns Hopkins University Press, 1982).

43. Keogh, *Wardian Case*, 208–217.

44. "Air conditioning is destiny. . . . As soon as air supply ceases to be an unproblematic premise of life processes and enters its technological stage, even this oldest pneumatic and atmospheric basic condition of human existence will have reached the threshold of modernity. From that point on, air mixtures and atmospheres will become objects of explicit production. . . . The bright side of the Enlightenment will become atmotechnics." Peter Sloterdijk, *Globes. Spheres II: Macrospherology*, trans. Wieland Hoban (South Pasadena, CA: Semiotext(e), 2014), 964–965.

45. The coincidence of plants and enslaved humans traveling through the same routes is a topic that we cannot deal in detail in this book. As Kijan Espahangizi puts it, in the context on his work on glass, "there is a genealogically older layer in which the glass vessel forms a matrix, literally a womb, in which the alchemist / modern man gives birth to a second, i.e. a new artificial nature"; Kijan Espahangizi, "The Twofold Section of Laboratory Glassware," in *Membranes Surfaces Boundaries. Interstices in the History of Science, Technology and Culture*, ed. Mathias Grote and Max Stadler (Berlin: Max Planck Institute for the History of Science, 2011), 27–44, at 28, In addition to the glass case, the slave ship was an ever more important vessel involved in the project of "a new artificial nature." As Christina Sharpe has emphasized in her critique of Sekula and Burch's

take on logistics, *The Forgotten Space*, the cargo ship echoes the "so-called migrant ship," "the prison," the slave ship, and therefore "the womb that produces blackness"; Christina Sharpe, *In the Wake: On Blackness and Being* (Durham, NC: Duke University Press, 2016), 27.

46. See Matthew Fuller, *Media Ecologies: Materialist Energies in Art Technoculture* (Cambridge, MA: MIT Press, 2005). Alexander Klose, *The Container Principle How a Box Changes the Way We Think*, trans. Charles Malcrum II (Cambridge, MA: MIT Press, 2015).

47. Kliment A. Timiriazev, "Croonian Lecture: The Cosmical Function of the Green Plant," *Proceedings of the Royal Society of London* 72, nos. 477–486 (1904): 424–461, at 461.

48. William R. Newman, *Newton the Alchemist: Science, Enigma, and the Quest for Nature's "Secret Fire"* (Princeton, NJ: Princeton University Press, 2018)

49. Timiriazev, "Croonian Lecture," 461.

50. Timiriazev, "Croonian Lecture," 424.

51. This experiment is for instance explained in Charles Bonnet, *Recherches sur l'usage des Feuilles dans les Plantes, et sur Quelques Autres Sujets Relatif à l'Histoire de la Végétation* (Göttingen: Elie Luzac, Fils, 1754).

52. See Jan van Ingenhousz, *Experiments upon Vegetables: Discovering Their Great Power of Purifying the Common Air in the Sun-Shine, and of Injuring It in the Shade and at Night. To Which Is Joined, a New Method of Examining the Accurate Degree of Salubrity of the Atmosphere* (London: P. Elmsly and H. Payne, 1779).

53. See Jean Sénebier, *Mémoires physico-chimiques sur l'influence de la lumière solaire pour modifier les êtres des trois règnes de la nature, et sur-tout ceux du règne végétal* (Geneva: Barthelemi Chirol, 1782), 75.

54. A succinct account of Aristotle's thoughts on the relation between plants, color and sunlight can be read in Elliot Weier, "The Structure of the Chloroplast," *Botanical Review* 4, no. 9 (1938): 497–530, at 497: "Of purely historical interest is the fact that Aristotle attempted to account for the green coloring of plants and apparently appreciated the importance of light in the production of the pigment, although one may be reading present day knowledge into his words. He states that yellow is derived from fire, white from air, water and the earth. Black develops when the elements are transformed from one to another. Green is a compound color and develops from a mixture of yellow (light) and black (the transformation of one element into another; air and water into earth). All plants developed a green color when they grew in the earth. They became white when the fire within them did not mix with the sun's rays."

55. Sénebier, *Mémoires physico-chimiques*, 200–202.

56. For a discussion on volumes and volumetry, see Jara Rocha and Femke Snelting, eds., *Volumetric Regimes: Material Cultures of Quantified Presence* (London: Open Humanities, 2022).

57. The botanist Matthias J. Schleiden explained this recursiveness when describing the mechanism of the microscope to the broad audience of his popular

science book, *The Plant: A Biography*. There he exposes first the features of Anton van Leeuwenhoek's simple microscope. It was a handheld device similar to a loupe, whose optics consisted only of a tiny sphere of glass. The principle of the compound microscope, he continues, "depends on a combination of the camera obscura with the simple microscope." That is, the amplified image after the first sphere would be internally projected and observed in recursion through a second sphere. See his full explanation in Matthias J. Schleiden, *The Plant: A Biography*, trans. Arthur Henfrey (London: Hippolyte Bailliere, 1848), 28.

58. Hugo von Mohl, *Untersuchungen Über Die Anatomischen Verhältnisse des Chlorophylls: Eine Inaugural-Dissertation* (Tübingen: Bähr, 1837), is often credited as the first work to describe the chloroplasts, or *Chlorophyllkörner*. However, botanists since the eighteenth century, such as Andrea Comparetti, Christian K. Sprengel, and Ludolph C. Treviranus, had described them, before von Mohl; see for instance, Conway Zirkle, "The Structure of the Chloroplast in Certain Higher Plants," *American Journal of Botany* 13, no. 5 (1926): 301–320, at 301. Even seventeenth-century microscopist Anton van Leeuwenhoek wrote about them; see Elliot Weier, "The Structure of the Chloroplast," *Botanical Review* 4, no. 9 (1938): 497–530, at 498.

59. Andreas F. W. Schimper, "Uber die Entwicklung der Chlorophyllkörner und Farbkörner," *Botanische Zeitung* 41 (1883): 105–114, at 112, quoted in Jan Sapp, Francisco Carrapiço, and Mikhail Zolotonosov, "Symbiogenesis: The Hidden Face of Constantin Merezhkowsky," *History and Philosophy of the Life Sciences* 24, no. 3/4 (2002): 413–440, at 419. Anecdotally, as it is discussed in the article by Sapp, Carrapiço, and Zolotonosov, these early insights on the possibility of symbiogenesis included the important contribution of the otherwise infamous Constantin Merezhkowsky, often credited wrongly as an earlier proponent of the idea. Years later, in the 1920s, Boris Kozo-Polyansky formulated a Darwinian theory of symbiogenesis. The modern endosymbiosis theory that explains how chloroplasts developed from cyanobacteria was developed by Lynn Margulis in her 1967 text "On the Origin of Mitosing Cells," *Journal of Theoretical Biology* 14, no. 3 (1967): 225–274.

60. See Espahangizi, "Wissenschaft im Glas." In addition to Espahangizi's work, the role of glass and bubbles in the interplay of scales has been highlighted as an important part of the material culture of natural sciences during the nineteenth century. Part of Simon Schaffer's long-time interest in glass (see the already mentioned Shapin and Schaffer, *Leviathan and the Air-Pump*, or Schaffer, "Glass Works: Newton's Prisms and the Uses of Experiment," in *The Uses of Experiment: Studies in the Natural Sciences*, ed. David Gooding, Trevor Pinch, and Simon Schaffer (Cambridge: Cambridge University Press, 1985): 67–104), has also included how bubbles came to be part of the modelling apparatus of natural sciences with scientists such as Michael Faraday and later on William Thompson using bubbles to trace and visualize fluid dynamics. See Simon Schaffer, "A Science Whose Business Is Bursting: Soap Bubbles as Commodities in Classical Physics," in *Things that Talk: Object Lessons from Art and Science*, ed. Lorraine Daston (New York: Zone, 2004), 147–194. On the role of glass as a modeling

material, see Lorraine Daston, "The Glass Flowers," in Daston, *Things that Talk*, 223–256.

61. Espahangizi, "Twofold Section of Laboratory Glassware."

62. Kijan Espahangizi, "Science in Glass: Material Pathologies in Laboratory Research, Glassware Standardization, and the (Un)Natural History of a Modern Material, 1900s–1930s," *Isis* 113, no. 2 (June 2022): 221–244.

63. Espahangizi, "Twofold Section of Laboratory Glassware," 36.

64. The entangled lines of genealogical investigation that come together in this chapter—and the whole book—concern various moments when techniques of light, space, and data unite, which was very much the case in the spheres of plant life and colonialism. This is the other aspect of plants as planetary agents—not only capturing and using the solar energy cast down from the skies but also moving across the surface as part of the established routes within Europe and between Europe and what were called New Worlds, or Neoeuropes, that were refabricated through introduction of imported plants and animals and which led to such disastrous practices as plantations. See Alfred W. Crosby, *Ecological Imperialism: The Biological Expansion of Europe, 900–1900* (Cambridge: Cambridge University Press, 2004).

65. Giuliana Bruno, *Surface: Matters of Aesthetics, Materiality, and Media* (Chicago: University of Chicago Press, 2014), 56.

66. Bruno, *Surface*, 5.

67. See also Anne Friedberg, *Window Shopping. Cinema and the Postmodern* (Berkeley: University of California Press, 1994).

68. "Die Erdoberfläche würde sich sehr verändern, wenn überall die Backsteinarchitektur von der Glasarchitektur verdrängt würde. / Es wäre so, als umkleidete sich die Erde mit einem Brillanten- und Emailschmuck." Scheerbart, *Glasarchitektur*, 21.

69. Janet Janzen, *Media, Modernity and Dynamic Plants in Early 20th Century German Culture* (Leiden: Brill, 2016), 92. This same thesis is also supported by Esther Leslie in *Synthetic Worlds: Nature, Art and the Chemical Industry* (London: Reaktion, 2006), 108. See also the 1957 novel *Gläserne Bienen* by Ernst Jünger; *The Glass Bees*, trans. Louise Bogan and Elizabeth Mayer (New York: New York Review Books, 2000).

70. Richard Drayton, *Nature's Government: Science, Imperial Britain, and the "Improvement" of the World* (New Haven, CT; London: Yale University Press, 2000).

71. See Beatriz Colomina, "Enclosed by Images: The Eameses' Multimedia Architecture," *Grey Room* 2 (Winter 2001): 7–29.

72. This point relates to the fascinating example of artificial light environments for plants in cases such as phytotron research centers of closed chambers and "sun simulators," used to investigate effects of radiation, spectral distribution, and UV-C, UV-B, and UV-A light. Stephan Thiel, Thorsten Döhring, Matthias Köfferlein, Andre Kosak, Peter Martin, and Harald K. Seidlitz, "A Phytotron

for Plant Stress Research: How Far Can Artificial Lighting Compare to Natural Sunlight?," *Journal of Plant Physiology* 148, no. 3–4 (1996): 456–463. The idea of thermocontrol and weather production was a persistent part of the discursive imaginaries these spaces created by operative means with fluorescent lights and air conditioning systems: "Light, temperature, humidity, gas content of the air, wind, rain, and fog—all these factors can be simultaneously and independently controlled," as the first phytotron facility at Caltech's Earhart Plant Research Laboratory was described by its founding member, plant physiologist Frits W. Went in 1949. Frits W. Went, "The Phytotron: Caltech Dedicates Its Fabulous Weather Factory," *Engineering and Science* 12, no. 9 (June 1949): 3. On the history of phytotrons, see David P. D. Munns, *Engineering the Environment: Phytotrons and the Quest for Climate Control in the Cold War* (Pittsburgh: University of Pittsburgh Press, 2017). On thermocontrol and elemental media, see Yuriko Furuhata, *Climactic Media: Transpacific Experiments in Atmospheric Control* (Durham, NC: Duke University Press, 2022).

73. "En effet, vient-on a transporter leur image dans l'appareil de M. Daguerre, ces parties vertes ne s'y trouvent pas reproduites, comme si tous les rayons chimiques, essentiels aux phenomenes daguerriens, avaient disparu dans la feuille, absorbés et retenus par elle." Jean-Baptiste Dumas and Jean-Baptiste Boussingault, *Essai de statique chimique des étres organisés* (Paris: Fortin, Masson, 1842), 24–25.

74. See note 57.

75. Olaf Breidbach, "Representation of the Microcosm: The Claim for Objectivity in 19th Century Scientific Microphotography," *Journal of the History of Biology* 35, no. 2 (2002): 221–250.

76. "To obtain such high magnifications one has to follow a special procedure: A photo of a certain preparation should be produced. This photo should be placed under a microscope and a second photo should be produced. To enhance the magnification of the original, a microphotograph of this second photo could be done. According to the authors, the resulting photos showed magnifications between 8000× and 30,000×. These shots revealed structural peculiarities which were not discernable by direct observation. These authors used the technique in such a way that it was not the microscopic preparations, but the photographs which were thought to provide accurate information about the structure of the microcosm." Breidbach, "Representation of the Microcosm," 232.

77. See Howard Caygill, "Harold Wager and the Photography of Plants," *Photographies* 14, no. 3 (2021): 505–519.

78. Kaja Silverman, *The Miracle of Analogy, or The History of Photography, Part 1* (Stanford, CA: Stanford University Press, 2015), 85.

CHAPTER 2

1. Giacomo Ciamician, "The Photochemistry of the Future," *Science* 36, no. 926 (1912): 385–394.

2. Elizabeth DeLoughrey and George Handley, *Postcolonial Ecologies: Literatures of the Environment* (New York: Oxford University Press, 2011), 11; Luke Keogh, *The Wardian Case: How a Simple Box Moved Plants and Changed the World* (Chicago: University of Chicago Press, 2020).
3. For a discussion about this obscuring of vegetal life in Western philosophy, see Michael Marder, *Plant-Thinking: A Philosophy of Vegetal Life* (New York: Columbia University Press, 2013), 19–36.
4. On the importance of herbaria and botanical illustration among these practices and treatises, see Londa Schiebinger, *Plants and Empire: Colonial Bioprospecting in the Atlantic World* (Cambridge, MA: Harvard University Press, 2007), and Daniella Bleichmar, *Visible Empire: Botanical Expeditions and Visual Culture in the Hispanic Enlightenment* (Chicago: University of Chicago Press, 2012). We do not approach in more detail these practices in the book as we want to focus on nonrepresentational photographic approaches to plants and territories.
5. On roots as pulled down by gravity, see Soraya de Chadarevian, "Laboratory Science versus Country-House Experiments: The Controversy between Julius Sachs and Charles Darwin," *British Journal for the History of Science* 29, no. 1 (1996): 17–41, at 24. On the hypothesis of heat bending plants, see Craig W. Whippo and Roger P. Hangarter, "Phototropism: Bending towards Enlightenment," *Plant Cell* 18, no. 5 (2006): 1110–1119, at 1111. On the irritability of the mimosa in mechanical terms, see Julius Sachs, *Lectures on the Physiology of Plants* (Oxford: Clarendon Press, 1887), 646–648. The *Mimosa pudica*, one of the best-known examples of a plant with retractile leaves, was added to the Linnean taxonomy in 1753.
6. Joseph Reynolds Green, *A History of Botany, 1860–1900: Being a Continuation of Sachs History of Botany, 1530–1860* (Oxford: Clarendon Press, 1909), 458–459. Not by chance, decades later, Charles Darwin and his son Francis carried on a successful series of observations that demonstrated the existence of signal transmission within the plants.
7. On the differences between technoscientific cultures regarding the discussion about plant perception, see de Chadarevian, "Laboratory Science versus Country-House Experiments." On the development of buildings conceived as laboratories and controlled environments, see Christian Reiß "The Biologische Versuchsanstalt as a Techno-Natural Assemblage: Artificial Environments, Animal Husbandry, and the Challenges of Experimental Biology," in *Vivarium: Experimental, Quantitative, and Theoretical Biology at Vienna's Biologische Versuchsanstalt*, ed. Gerd B. Müller (Cambridge, MA: MIT Press, 2017), 115–132. On the transformation of the German biological sciences into quantitative disciplines relying on precision laboratories, and the importance of the development on the cell theory in this process, see Timothy Lenoir, *The Strategy of Life: Teleology and Mechanics in Nineteenth-Century German Biology* (Chicago: University of Chicago Press, 1989), 112–155. On the differences with the Austro-Hungarian context, see Deborah R. Coen, *Vienna in the Age of Uncertainty: Science, Liberalism, and Private Life* (Chicago: University of Chicago Press, 2007), 96–104. For a broader view of the development

of the technological complex of life sciences that preceded this decade, see Olga Elina, Susanne Heim, and Nils Roll-Hansen, "Plant Breeding on the Front: Imperialism, War, and Exploitation," *Osiris* 20 (2005): 161–179.

8. Schmidgen Henning, *The Helmholtz Curves: Tracing Lost Time*, trans. Nils F. Schott (New York: Fordham University Press, 2014).
9. Soraya de Chadarevian, "Graphical Method and Discipline: Self-Recording Instruments in Nineteenth-Century Physiology," *Studies in History and Philosophy of Science* 24 (1993): 267–291. On the importance of Pfeffer's experiments in the observation of the relation between plant stimuli and movements, see Green, *History of Botany*, 111–112.
10. On the importance of autographic devices in the study of plant movements—such as the systematic use by Charles Darwin and his son Francis of smoked glass-plates to register periodic movements—see Oliver Gaycken "The Secret Life of Plants: Visualizing Vegetative Movement, 1880–1903," *Early Popular Visual Culture* 10, no. 1 (2012): 51–69.
11. On the early use in the context of botany of cameraless photography, such as the well-known prints by Anna Atkins, see Katharina Steidl, "Leaf Prints: Early Cameraless Photography and Botany," *PhotoResearcher* 17 (2012): 26–35.
12. On photography and the telescope—an example is John Adams Whipple, *The Moon*, 1857–1860—see Jai McKenzie, *Light and Photomedia: A New History and Future of the Photographic Image* (London: I. B. Tauris, 2014), 25. See the early experiments with microscope daguerrotypes of Alfred Donné and Leon Foucault, published in 1844, discussed in Lorraine J. Daston and Peter Galison, *Objectivity* (New York: Zone, 2007), 134. Already in 1842 John William Draper had made daguerreotypes of solar spectra; see Klaus Hentschel, *Mapping the Spectrum: Techniques of Visual Representation in Research and Teaching* (Oxford: Oxford University Press, 2002), 197.
13. Scott Curtis, "Photography and Medical Observation," in *The Educated Eye: Visual Culture and Pedagogy in the Life Sciences*, ed. Nancy Anderson and Michael R. Dietrich (Hanover, NH: Dartmouth College Press, 2012), 68–93, at 68. See Daston and Galison, *Objectivity*, 120.
14. Curtis, *Photography and Medical Observation*, 76–78. For a cultural and material approach to the use of photographs in nineteenth-century botany, see Caroline Fieschi, *Photographier les plantes au XIXe siècle: La photographie dans les livres de botanique* (Paris: CTHS Sciences, 2008).
15. The experiments discussed in this chapter are contemporary with the projection apparatuses designed to create moving images of living things, which proliferated in the visual instruction of life sciences. As Schmidgen explains, these media devices that the German physiologist and pharmacologist Carl Jacobj named *Anordungs* (arrangements), were "rather complex assemblages of organic and mechanical parts, electrical current and light rays, lenses, carbon rods and mirrors, frogs, wooden boards, and water jars." See Henning Schmidgen, "Cinematography without Film: Architectures and Technologies of Visual

Instruction in Biology around 1900," in Anderson and Dietrich, *Educated Eye*, 94–120, at 99. These instruments include, for instance: the Universal Projection Apparatus by Leitz; the Epidiascope by Zeiss in Jena and the Universal Projectoscope by Stoelting in Chicago; and episcopic projectors of Salomon Stricker (1834–1898) in Vienna and Carl Kaiserling (1869–1942) in Berlin. Johann Nepomuk Czermak's institute (1873) in Leipzig was named Spectatorium. For more details, see Schmidgen, "Cinematography without Film."

16. Raoul H. Francé, *Germs of Mind in Plants* (Chicago: Kerr, 1905), 17. On his later work on the topic of bionics and "functional surfaces," see Jan Mueggenburg, "Clean by Nature: Lively Surfaces and the Holistic-Systemic Heritage of Contemporary Bionik," *Communication+1* 3, no. 1 (2014): 1–28. On Francé's use of photographic sequences to track vegetal movements, see Matthew Vollgraff, "Gestes végétaux," in *Puissance du végétal et cinéma animiste: La vitalité révelée par la technique*, ed. Teresa Castro, Perig Pitrou, and Parie Rebecchi (Paris: Les Presses du Réel, 2020), 195–223.

17. With August Pauly and Adolf Wagner, Francé formed a group of "psychovitalists" which predicated an animist approach to nature. See Vollgraff, "Gestes végétaux," 201–204. Also in Vienna, other forms of "mechanistic vitalisms" were practiced, in the by then well-known and respected Biologische Versuchsanstalt (Institute of Biology), the Vivarium. See Gerd B. Müller, "Biologische Versuchsanstalt: An Experiment in the Experimental Sciences," in Müller, *Vivarium*, 3–18. On Hans Driesch and Henri Bergson, see Jane Bennett, *Vibrant Matter: A Political Ecology of Things* (Durham, NC: Duke University Press, 2010), 62–81. On the disputes against vitalism in the long German nineteenth century, see Lenoir, *Strategy of Life*.

18. Inga Pollmann, *Cinematic Vitalism: Film Theory and the Question of Life* (Amsterdam: Amsterdam University Press, 2018), 27–34.

19. See Janet Janzen, *Media, Modernity and Dynamic Plants in Early 20th Century German Culture* (Leiden: Brill, 2016) as well as the recently edited collection on this topic, Castro, Pitrou, and Rebecchi, *Puissance du végétal et cinéma animiste*. A filmography would include Germaine Dulac, Max Reichmann, Jean Comandon, and Friedrich W. Murnau, among others. A shared argument in the collection edited by Castro, Pitrou, and Rebecchi is that this vegetal animism can also be traced in some experimental practices of the first decades of the twenty-first century. This theme also relates to Beny Wagner's practice-based PhD project on cinema and metabolism at the Winchester School of Art, in which he outlines early cinema as an epistemological staging of questions of metabolism; this also feeds into his work on moving image aesthetics.

20. Howard Caygill, "Harold Wager and the Photography of Plants," *Photographies* 14, no. 3 (2021): 505–519, at 515.

21. Josef Maria Eder, *History of Photography*, trans. Edward Epstean (Toronto: Dover, 1978), 417.

22. Others in the plant science community included Sachs in Berlin, and Pfeffer in Leipzig. See de Chadarevian, "Laboratory Science versus Country-House Experiments."

23. See, for instance, his publication of technical microscopy: Julius Wiesner, *Einleitung in die technische Mikroskopie nebst mikroskopisch-technischen Untersuchungen* (Vienna: W. Braumüller, 1867). He extended his practice to the microscopic analysis of paper—including old manuscripts—while keeping a relation with the paper industry, see Anna-Grethe Rischel, "Julius von Wiesner and His Importance for Scientific Research and Analysis of Paper," *Paper History* 18, no. 1 (2014): 31–38. He dealt also with textiles; see Susanne Heim, Carola Sachse, and Mark Walker, *The Kaiser Wilhelm Society under National Socialism* (Cambridge: Cambridge University Press, 2009). His views about the applications of plant physiology to agriculture and other domains can be read in his inaugural lecture, Julius Wiesner, "Die Beziehungen der Pflanzenphysiologie zu den anderen Wissenschaften," in *Inaugurationsrede im Festsaale der Universität* (Vienna: Universität Wien, 1898).

24. On Wiesner as a promoter of a universal knowledge, see Coen, *Vienna in the Age of Uncertainty*, 164–165. His relation with the parliament is mentioned by Kärin Nickelsen, *Explaining Photosynthesis: Models of Biochemical Mechanisms, 1840–1960* (Dordrecht: Springer, 2019), 182. Other colleagues, such as Jose Maria Eder, exhibited regularly their well-known photographic work with experimental techniques, such as x-rays.

25. Julius Wiesner, *Der Lichtgenuss der Pflanzen: Photometrische und physiologische Untersuchungen mit besonderer Rücksichtnahme auf Lebensweise, geographische Verbreitung und Kultur der Pflanzen* (Leipzig: W. Engelmann, 1907), 251.

26. His work on the growth of plants at a microscopic scale is gathered in Julius Wiesner, *Die Elementarstructur und das Wachstum der lebenden Substanz* (Vienna: A. Hölder, 1892) where he proposed his theory of the plasoms as the particles that made up vegetable matter.

27. Wiesner, *Der Lichtgenuss der Pflanzen*, 68.

28. In his work on the history of photography, Eder tracks the development of this invention and explains how the design by Bunsen and Roscoe improved the one by F. J. Malaguti in 1839; see Eder, *History of Photography*, 415. On photometers, see also Michael Pritchard, "Actinometers and Exposure Measurement," in *Encyclopedia of Nineteenth-Century Photography*, ed. John Hannavy (London: Routledge, 2013), 4–5.

29. Wiesner, *Der Lichtgenuss der Pflanzen*, 14.

30. This was achieved through the notion of the relative *Lichtgenuss*: with the aid of assistants, several papers were exposed to light for the same amount of time. One by one, the blackened papers were introduced into the Insolator and pulled out, together with the measuring photographic paper, in order to measure the time the test paper needed to acquire each of the colors. These relative times could then be easily compared and averaged against values of daylight darkening to become absolute figures.

31. Wiesner, *Der Lichtgenuss der Pflanzen*, 2.

32. As Eder puts it, "exposure meters with silver salt papers and normal gray tints with tables were introduced by Stanley (1886), Wynne (1893), Alfred Watkins

('Standard Exposure Meter') 1890, (W. G.) Watkins ('Beemeter'), and others"; Eder, *History of Photography*, 449. For more details, see Pritchard, "Actinometers and Exposure Measurement."

33. Kliment A. Timiriazev, *The Life of the Plant: Ten Popular Lectures* (Moscow: Foreign Languages Publishing House, 1958), 414.

34. Neovitalism's main contributor, Hans Driesch, defined Wiesner as a "static teleologist," in *Hans Driesch, Geschichte des Vitalismus* (Leipzig: Johann Ambrosius Barth, 1922), 163. See also Oskar Ewald, "Die deutsche Philosophie im Jahre 1910," *Kant-Studien* 16, no. 1–3 (1911): 382–430, at 429.

35. Wiesner, *Die Beziehungen der Pflanzenphysiologie*, 67.

36. Wiesner, *Der Lichtgenuss der Pflanzen*, 107s

37. The habitus as the ensemble of adaptations and site-specific regulations of the plants is used throughout his text. Another metaphor used by Wiesner is that of the economy: the interweaving between the shape and the plant's surroundings is understood as an economy of light. See Wiesner, *Der Lichtgenuss der Pflanzen*, 71–74.

38. Lenoir, *Strategy of Life*, 76.

39. Vollgraff, "Gestes végétaux," 204.

40. In a letter sent to biologist George J. Romanes, Charles Darwin's writes: "Wiesner and Tieghem seem to think that this is explained by calling the whole process 'induction,' borrowing a term used by some physico-chemists (of whom I believe Roscoe is one), and implying an agency which does not produce any effect for some time, and continues its effect for some time after the cause has ceased. I believe (?) that photographic paper is an instance. I must ask Leonard [his son] whether an interrupted light acts on it in the same manner as on a plant." In a letter the following day, Darwin expressed his concern about the fact that "most botanists believe that light causes a plant to bend to it in as direct a manner as light affects nitrate of silver." *The Life and Letters of George John Romanes*, edited by Ethel D. Romanes (London: Longmans, 1908), 119. Wiesner's experiment took place in 1878. "The inductive nature of the response was finally confirmed when Julius von Wiesner (1838–1919) showed that plants continue to bend toward a light source even after the light is turned off," Whippo and Hangarter, "Phototropism," 1111. Photochemical induction was introduced as a phenomenon linked to the formation of images in photography: "The laws of photo-chemical induction which we have developed in this Part, explain most completely many of the singular phenomena which lie at the foundation of the photographic processes." Robert Bunsen and Henry Enfield Roscoe, "Photo-Chemical Researches. Part II: Phenomena of Photo-Chemical Induction," *Philosophical Transactions of the Royal Society of London* 147 (1857): 381–402, at 400.

41. Luce Lebart, "La Photographie d'Origine Végétale," in Castro, Pitrou, and Rebecchi, *Puissance du Végétal et Cinéma Animiste*, 119–127.

42. Lebart, "Photographie d'Origine Végétale." On Timiriazev's prints, see Kliment A. Timiriazev, "Croonian Lecture: The Cosmical Function of the Green Plant," *Proceedings of the Royal Society of London* 72, nos. 477–486 (1904): 424–461.
43. The terms refer to whether the growth and orientation of leaves occurred facing the direction of the incident light rays, independent of it, completely perpendicular to it, or facing only the most intense light, respectively. See Wiesner, *Der Lichtgenuss der Pflanzen*, 70–74.
44. Caygill, "Harold Wager and the Photography of Plants," 515.
45. Timiriazev, *Life of the Plant*, 180. The metaphor of forests as factories has evolved into a speculative model of a biosynthetic industrial future, see chapter 7.
46. In relation to exposure tables, see Pritchard, "Actinometers and Exposure Measurement."
47. John Tresch, *The Romantic Machine: Utopian Science and Technology after Napoleon* (Chicago: University of Chicago Press, 2012), 116.
48. Tresch, *Romantic Machine*, 115. See more generally pages 89–122.
49. Such an expanded focus on operational images is described in Jussi Parikka, *Operational Images: From the Visual to the Invisual* (Minneapolis: University of Minnesota Press, 2023). See also Tomáš Dvořák and Jussi Parikka, "Measuring Photographs," *Photographies* 14, no. 3 (2021): 443–457.
50. Vered Maimon, "On the Singularity of Early Photography: William Henry Fox Talbot's Botanical Images," *Art History* 34, no. 5 (2011): 958–977, at 963. See also Michelle Henning, *Photography: The Unfettered Image* (London: Routledge, 2018).
51. Maimon, "On the Singularity of Early Photography," 967.
52. Thomas Macho, "Second-Order Animals: Cultural Techniques of Identity and Identification," *Theory, Culture & Society* 30, no. 6 (2013): 30–47, at 31.
53. Bernhard Siegert, *Cultural Techniques: Grids, Filters, Doors, and Other Articulations of the Real*, trans. Geoffrey Winthrop-Young (New York: Fordham University Press, 2015), 13.
54. Dvořák and Parikka, "Measuring Photographs." See also Parikka, *Operational Images*.
55. Other visually significant organs of plants, such as the flowers, explore a broader spectrum of color-encoded processes of exchange of matter and energy in the planet, that goes "beyond green"; see *Prismatic Ecology: Ecotheory beyond Green*, ed. Jeffrey J. Cohen (Minneapolis: University of Minnesota Press, 2013). See also important elaborations and critiques of the discourses of "greenness" in the project "GREEN Revisited: Encountering Emerging Naturecultures" with various workshops, exhibitions, and publications over the past years. http://green.rixc.org/.
56. Michael Marder, *Plant-Thinking: A Philosophy of Vegetal Life* (New York: Columbia University Press, 2013), 81.
57. Emanuele Coccia, *The Life of Plants: A Metaphysics of Mixture* (Cambridge: Polity, 2018), 5.

58. Wiesner, *Der Lichtgenuss der Pflanzen*, 96.

59. Georges Bataille, *Visions of Excess: Selected Writings, 1927–1939* (Minneapolis: University of Minnesota Press, 1985), 3.

60. The calculation of the LAI is slightly different, however: it is the sum of the one-sided areas of leaves in relation to the area of the ground.

61. Vladimir I. Vernadsky, *The Biosphere* (New York: Copernicus, 1998), 78.

62. Rolf G. Kuehni, *Color Ordered: A Survey of Color Order Systems from Antiquity to the Present* (Oxford: Oxford University Press, 2008), 86.

63. Wiesner, *Der Lichtgenuss der Pflanzen*, 220. The Insolator, the Lichtfläche, and the use of the color charts all involve the presence of devices that belong to what Wolfgang Ernst calls "measuring media": Wolfgang Ernst, *Digital Memory and the Archive*, ed. Jussi Parikka (Minneapolis: University of Minnesota Press, 2013), 178. Such devices enabled the observer to distinguish previously unperceived natural phenomena. Wiesner's image-plants could be seen beyond the threshold of the naked eye's capacity, in a way similar to what was happening in the chronophotographic experiments of Eadweard Muybridge, Ottomar Anschütz, Étienne-Jules Marey and others. See McKenzie, *Light and Photomedia*, 32–33.

64. Wiesner, *Der Lichtgenuss der Pflanzen*, 69

65. "Am 30. März 1893 um 10h45m a.m. beobachtete ich im Wiener Augarten eine Intensität des gesamten Tageslichtes = 0.427. Am Südostrande eines dort befindlichen, dichten, noch gänzlich unbelaubten, aus hochstämmigen Bäumen zusammengesetzten Roßkastanienbestandes herrschte aber im vollen Sonnenlichte gleichzeitig bloß eine Intensität = 0.299. Im Schatten eines Roßkastanienstammes (NE) betrug die Intensität nur 0.023." Wiesner, *Der Lichtgenuss der Pflanzen*, 69.

66. Carl E. Schorske, *Fin-De-Siecle Vienna: Politics and Culture* (Cambridge: Cambridge University Press, 1985), 322.

67. Robert Musil, *The Man without Qualities*, trans. Sophie Wilkins and Burton Pike (London: Picador, 2011).

68. Coen, *Vienna in the Age of Uncertainty*, 101.

69. Here Coen is using Carlo Ginzburg's terms, see Coen, *Vienna in the Age of Uncertainty*, 102. In chapter 5 we address Carlo Ginzburg's historically situated "evidential paradigm" in the analysis of aerial images. See Carlo Ginzburg, *Clues, Myths, and the Historical Method*, trans. John Tedeschi and Anne C. Tedeschi (Baltimore, MD: Johns Hopkins University Press, 2013).

70. It was a complex circuit of mechanical, electrical and chemical components. see Erwin Bunning, *Ahead of His Time: Wilhelm Pfeffer, Early Advances in Plant Biology* (Ottawa: Carleton University Press, 1989), 66.

71. Coen, *Vienna in the Age of Uncertainty*, 101.

72. As film scholar Oliver Gaycken has observed, there was an important tradition of time-lapse media in plant sciences before the photographic ones. Charles

Darwin and his son Francis used systematically smoked glass-plates and drawings to register periodic movements. Gaycken, "Secret Life of Plants," 56.

73. Gaycken, "Secret Life of Plants," 58. Pfeffer's fondness of technical imaging found applications in many domains: "In the winter semester of 1919–20, we saw the microscopic projections which Pfeffer had already described in the year 1900; the protoplasmic streaming, the plasmolysis, and the formation and growth of precipitation membranes. Examples of projections in silhouette were the stimulation of tendrils, the thermonastic opening of tulip buds, and the formation of oxygen bubbles—very impressive—with Elodea. In the latter case, it was demonstrated that the magnitude of the stream of bubbles depended on the light intensity (a diffuse light was obtained by using a wire mesh). Pfeffer also presented time-lapse photos of the following subjects: the geotropic raising to an upright position of a plant of Impatiens glandulifera, the sleep movements of Desmodium gyrans, the movements of Mimosa spegazzinii, and the germination of Vicia faba. He also demonstrated the nutation of the pedicel, and the blossoming and withering of tulips." Bunning, *Ahead of His Time*, 120.

74. See Gaycken, "Secret Life of Plants."

75. The German film pioneer Oskar Messner helped popularize them by showing them at film evenings before main features. Janzen, *Media, Modernity and Dynamic Plants*, 2.

76. Schmidgen, "Cinematography without Film," 114. On visuality and technosciences, see also the classic text by Donna Haraway, *Simians, Cyborgs, and Women: The Reinvention of Nature* (New York: Routledge, 1991), 184–201.

77. Joanna Zylinska, *Nonhuman Photography: Theories, Histories, Genres* (Cambridge, MA: MIT Press, 2017), 17.

78. Sometimes spelled as klinostat.

79. De Chadarevian, "Laboratory Science versus Country-House Experiments," 39. Sachs had additionally been Pfeffer's mentor. Bunning, *Ahead of His Time*, 18.

80. Wilhelm Pfeffer, *Pflanzenphysiologie; Ein Handbuch der Lehre vom Stoffwechsels und Kraftwechsels in der Pflanze* (Leipzig: W. Engelmann, 1897), 569.

81. Coccia, *Life of Plants*, 5.

82. See Keller Easterling, *Medium Design: Knowing How to Work on the World* (London: Verso 2020).

83. Rina Scott, "On the Movements of the Flowers of Sparmannia Africana, and Their Demonstration by Means of the Kinematograph," *Annals of Botany* 17, no. 68 (1903): 761–77, at 773.

84. Scott, "On the Movements of the Flowers," 772.

85. The kammatograph, invented by Leonard Kamm, was one of the early predecessors or the cinema camcorder, see "The Kammatograph," *Image* 1, no. 8 (November 1952): 3.

86. Scott, "On the Movements of the Flowers," 773. Wilhelm Pfeffer, "Die Anwendung des Projectionsapparates zur Demonstration von Lebensvorgängen,"

Jahrbüchen für Wissenschaftliche Botanik 35, no. 4 (1900): 711–745. On the use of the clinostat in the regulation of illumination, see Bunning, *Ahead of His Time*, 66.

87. Timiriazev, *Life of the Plant*, 68.
88. Harold Wager, "The Perception of Light in Plants," *Annals of Botany* 23, no. 91 (1909): 464.
89. This expression comes from Reiß, "Biologische Versuchsanstalt as a Techno-natural Assemblage."

CHAPTER 3

1. Giulia Rispoli and Jacques Grinevald, "Vladimir Vernadsky and the Co-Evolution of the Biosphere, the Noosphere, and the Technosphere," *Technosphere Magazine*, June 20, 2018, https://technosphere-magazine.hkw.de/p/Vladimir-Vernadsky-and-the-Co-evolution-of-the-Biosphere-the-Noosphere-and-the-Technosphere-nuJGbW9KPxrREPxXxz95hr. Wladimir I. Vernadsky, "The Biosphere and the Noösphere," *American Scientist* 33, no. 1 (1945): xxii, 1–12.
2. Hannah Holleman, *Dust Bowls of Empire: Imperialism, Environmental Politics* (New Haven, CT: Yale University Press, 2018).
3. Siegfried Giedion, *Mechanization Takes Command: A Contribution to Anonymous History* (New York: Oxford University Press, 1970), 142.
4. Deborah Coen, *Climate in Motion: Science, Empire, and the Problem of Scale* (Chicago: University of Chicago Press, 2018).
5. David Moon, "The Environmental History of the Russian Steppes: Vasilii Dokuchaev and the Harvest Failure of 1891," *Transactions of the Royal Historical Society* 15 (2005): 149–174, at 164.
6. Moon, "Environmental History of the Russian Steppes," 162–163.
7. Vernadsky, "Biosphere and the Noösphere," 5. On chemical modernity and surface aesthetics, see Esther Leslie, *Synthetic Worlds: Nature, Art and the Chemical Industry* (London: Reaktion, 2005).
8. Max Liboiron, *Pollution Is Colonialism* (Durham, NC: Duke University Press, 2021), 110.
9. Anna Lowenhaupt Tsing, "'On Nonscalability: The Living World Is Not Amenable to Precision-Nested Scales," *Common Knowledge* 18, no. 3 (2012): 505–524.
10. Tsing, "'On Nonscalability," 505.
11. Alfred J. Lotka, *Elements of Physical Biology* (Baltimore, MD: Williams and Wilkins, 1925), 360.
12. Liboiron, *Pollution Is Colonialism*, 84.
13. Bruno Latour, *Down to Earth: Politics in the New Climatic Regime*, trans. Catherine Porter (Cambridge: Polity, 2018), 78

14. See Dipesh Chakrabarty, "The Planet: An Emergent Humanist Category," *Critical Inquiry* 46, no. 1 (2019): 1–31.
15. Lynn Margulis and Dorion Sagan, *What Is Life?* (New York: Simon and Schuster, 1995), 47.
16. Vladimir I. Vernadsky, *The Biosphere*, trans. David B. Langmuir, rev. and annotated Mark A. S. McMenamin (New York: Copernicus, 1998), 88.
17. Giulia Rispoli, "Between 'Biosphere' and 'Gaia': Earth as a Living Organism in Soviet Geo-Ecology," *Cosmos and History* 10, no. 2 (2014): 78–91, at 80–82.
18. Examples would be *The Theory of the Earth* (1788) by Scottish geologist James Hutton (1726–1797), the series of works on the estimation of the age of Earth by Lord Kelvin (1824–1907), and *Worlds in the Making* (1908) by Swedish chemist Svante Arrhenius (1859–1927), where the greenhouse effect is proposed; M. D. H. Jones and Ann Henderson-Sellers, "History of the Greenhouse Effect," *Progress in Physical Geography* 14, no. 1 (1990): 1–18. For other related references in the fields of meteorology and geography, see Mike Davis, "The Coming Desert: Kropotkin, Mars and the Pulse of Asia," *New Left Review* 97 (January–February 2016), https://newleftreview.org/II/97/mike-davis-the-coming-desert.
19. Rispoli, "Between 'Biosphere' and 'Gaia,'" 80.
20. John Murray, "1899: The State of Ocean Science," *Scottish Geographical Magazine* 15 (1899): 505–522, https://oceanexplorer.noaa.gov/history/docs/science.html.
21. Jacques Grinevald, "Introduction: The Invisibility of the Vernadskian Revolution," in Vernadsky, *The Biosphere*, 23. See also Rispoli and Grinevald, "Vladimir Vernadsky and the Co-Evolution of the Biosphere."
22. Vernadsky, *The Biosphere*, 39.
23. Grinevald has observed that this wording is a specific reference to the work of French physiologist Claude Bernard, who characterized the living organism as an entity that kept a distinction between the internal and cosmic milieus. Grinevald, "Introduction," 29.
24. Vernadsky, *The Biosphere*, 39
25. Vernadsky was not able to respond to the question concerning the nature of the holistic mechanism: "I will not speculate here about the existence of the mechanism, but rather will observe that it corresponds to all the empirical facts and follows from scientific analysis"; Vernadsky, *The Biosphere*, 40. Such a statement is an example of the empirical generalizations we discuss above.
26. Vernadsky, *The Biosphere*, 51.
27. Vernadsky, *The Biosphere*, 40.
28. On different scales, see Joshua DiCaglio, *Scale Theory: A Nondisciplinary Inquiry* (Minneapolis: University of Minnesota Press, 2021), 9. For a relation between scale and theories of complex systems in the second half of the twentieth century, see Grinevald, "Introduction," 30.
29. Susan L. Brantley, Martin B. Goldhaber, and K. Vala Ragnarsdottir, "Crossing Disciplines and Scales to Understand the Critical Zone," *Elements* 3, no. 5 (2007):

307–314. Daniel DeB. Richter Jr. and Megan L. Mobley, "Monitoring Earth's Critical Zone," *Science* 326, no. 5956 (2009): 1067–1068. See also "The Critical Zone Collaborative Network" (CZNet), https://criticalzone.org/. The Critical Zone was the title and the focus of the exhibition at ZKM (Center for Art and Media Karlsruhe) (2020–2022) curated by Bruno Latour and Peter Weibel. See also the subsequent volume Bruno Latour and Peter Weibel, eds., *Critical Zones: The Science and Politics of Landing on Earth* (Cambridge, MA: MIT Press, 2020).

30. Frank Wigglesworth Clark, *The Data of Geochemistry*, United States Geological Survey Bulletin no. 330 (Washington, DC: Government Printing Office, 1908), 21
31. Vernadsky *The Biosphere*, 51–59.
32. Gabrielle Hecht, "Interscalar Vehicles for an African Anthropocene: On Waste, Temporality, and Violence," *Cultural Anthropology* 33, no. 1 (2018): 109–141.
33. Vernadsky *The Biosphere*, 44.
34. William H. F. Talbot, *The Pencil of Nature* (London: Longman, Brown, Green and Longmans, 1844), 4.
35. Vernadsky *The Biosphere*, 62.
36. Vernadsky, *The Biosphere*, 58.
37. Vernadsky, *The Biosphere*, 44.
38. Katherine N. Hayles, *How We Became Posthuman*. (Chicago: University of Chicago Press, 1999), 8.
39. Etienne Benson, *Surroundings: A History of Environments and Environmentalisms* (Chicago: University of Chicago Press, 2020), 113.
40. Benson, *Surroundings*, 115. Vernadsky writes: "In 1915, a 'Commission for the Study of the Productive Forces' of our country, the so-called KEPS, was formed at the Academy of Sciences. That commission, of which I was elected president, played a noticeable role in the critical period of the First World War. Entirely unexpectedly, in the midst of the war, it became clear to the Academy of Sciences that in Tsarist Russia there were no precise data concerning the now so-called strategic raw materials, and we had to collect and digest dispersed data rapidly to make up for the lacunae in our knowledge m. Unfortunately by the time of the beginning of the Second World War, only the most bureaucratic part of that commission, the so-called Council of the Productive Forces, was preserved, and it became necessary to restore its other parts in a hurry"; Vernadsky, "Biosphere and the Noösphere," 5.
41. Vernadsky, *The Biosphere*, 153.
42. Benson, *Surroundings*, 128.
43. Joan Martínez Alier, "Ecología Industrial y Metabolismo Socioeconómico: Concepto y Evolución Histórica," *Economía Industrial* 351 (2003): 15–26. Grinevald, "Introduction," 26. John Bellamy Foster, *Marx's Ecology: Materialism and Nature* (New York: New York University Press, 2000). To this effect, from a cultural studies of science perspective, Vernadsky's work has been addressed as a marker of a characteristic style of thinking whose ramifications can be traced in later

developments of Russian systems science and cybernetics; see Rispoli, "Between 'Biosphere' and 'Gaia.'"

44. Zachary Horton, *The Cosmic Zoom: Scale, Knowledge, and Mediation* (Chicago: University of Chicago Press, 2021).

45. Vernadsky, *The Biosphere*, 53.

46. Vernadsky quoted in Rispoli and Grinevald, "Vladimir Vernadsky and the Coevolution of the Biosphere." We have not, however, been able to locate the original quote.

47. The notion of a "planetary conveyor belt" features in some more recent discussions in speculative design and urbanism, such as architect Liam Young's work. Liam Young, "New City: Machines of Post Human Production," CCCB (Centre de Cultura Contemporània de Barcelona), December 16, 2015, https://www.cccb.org/en/multimedia/videos/liam-young-talks-about-new-city-machines-of-post-human-production/222592. See also Jussi Parikka, "Folds of Fashion: *Unravelled* and the Planetary Surface," in *Surface and Apparition: The Immateriality of Modern Surface*, ed. Yeseung Lee (London: Bloomsbury, 2020), 19–36.

48. Vernadsky, *The Biosphere*, 59.

49. Vernadsky, *The Biosphere*, 142; see also 16.

50. Vernadsky, *The Biosphere*, 143.

51. Horton, *Cosmic Zoom*, 49. In the cases explored by Horton this tension occurs between "a page or screen" in books or films such as Kees Boeke's *Cosmic View* and the Eames's *The Powers of Ten*, which place themselves in "a referential relationship with some other surface at a different scale" (45).

52. Horton, *Cosmic Zoom*, 48.

53. See Liboiron, *Pollution Is Colonialism*, 84.

54. Grinevald, "Introduction," 26.

55. Julius Wiesner, *Die Elementarstructur und das Wachstum der lebenden Substanz* (Vienna: A. Hölder, 1892).

56. Vernadsky, *The Biosphere*, 62.

57. Vernadsky, *The Biosphere*, 58.

58. Vernadsky, *The Biosphere*, 59.

59. Vernadsky, *The Biosphere*, 59.

60. Many of the parameters were defined in more detail of course, including how respiration was the ultimate determining factor for the minimum and maximum size of living bodies in contrast to physical constraints and the biosphere as limit cases for inert bodies. Vernadsky, "Biosphere and the Noösphere," 4.

61. Soraya de Chadarevian, "Laboratory Science versus Country-House Experiments: The Controversy between Julius Sachs and Charles Darwin," *British Journal for the History of Science* 29, no. 1 (1996): 17–41, at 24.

62. Vernadsky, *The Biosphere*, 59.

63. Vernadsky, *The Biosphere*.

64. See Kliment A. Timiriazev, "Croonian Lecture: The Cosmical Function of the Green Plant," *Proceedings of the Royal Society of London* 72, nos. 477–486 (1904): 424–61, at 439. Optical sensitizers are dyes that make silver halide crystals sensitive to certain light colors. Also, as cultural theorist Michelle Henning has remarked (*Photography: The Unfettered Image* (London: Routledge, 2018), 95–99), in the first decades of the twentieth century photography experienced its own particular revolution thanks, for instance, to the development of new and faster sensitizers. With their aniline dyes, corporations such as Bayer, AGFA, and BASF transferred their industrial mastery of the microtemporal synchronization of chemical cycles to shortened photographic exposure times, and by doing so expanded, in turn, the operational space of photography itself. Measurements and scientific practices relying on chronophotography, such as the ones linked to the experiments reviewed in chapter 2 are an example of this. Another clear case involves aerial photography: during the First World War, these dyes gave rise to specific sensitizers that "reshaped photography in response to the demands of aerial reconnaissance"; Henning, *Photography*, 99. In particular, sensitizers were developed to allow aerial cameras to see through the atmospheric haze.
65. Vernadsky, *The Biosphere*, 59.
66. This has been analyzed as a reason for the relative low impact of his work among his contemporaries; see Kiril M. Khailov, "Vladimir Ivanovich Vernadsky Originator of the Biosphere Concept," *La Mer* 32 (1994): 1–4, at 2. See also Rispoli, "Between 'Biosphere' and 'Gaia.'"
67. Vladimir I. Vernadsky, "The Evolution of Species and Living Matter: Appendix to the French Translation of *The Biosphere*" (1928), trans. Meghan Rouillard, *21st Century* (Spring–Summer 2012), 32–44, at 33; https://21sci-tech.com/Articles_2012/Spring-Summer_2012/05_Species_Matter.pdf. The text is from a speech originally given by Vernadsky to the Society of Naturalists of Leningrad on February 5, 1928, as per the translator's note (32).
68. Rispoli and Grinevald, "Vladimir Vernadsky and the Co-evolution of the Biosphere."
69. Vernadsky, "Evolution of Species and Living Matter," 44.
70. Vernadsky, "Evolution of Species and Living Matter," 38.
71. Vernadsky, "Evolution of Species and Living Matter," 41 On different strands of technics of nature, see Jussi Parikka, *Insect Media: An Archaeology of Animals and Technology* (Minneapolis: University of Minnesota Press, 2010).
72. Vernadsky quoted by Andrei V. Lapo, *Traces of Bygone Biospheres*, trans. V. Purto (Moscow: MIR, 1982)
73. Vernadsky illustrated his ideas about circulations and accumulations of living matter using a variety of data: from the British naturalist G. T. Carruthers, who had observed the annual flight of locusts over the Red Sea; from Nicolaus Steno, the acknowledgement that limestones were formed by skeletal remains of organisms; the formation of Dolomites as a result of the vital activity of

organisms; and the bacterial origin of the iron ores around lakes, described by C. G. Ehrenberg. See Lapo, *Traces of Bygone Biospheres.*

74. Lapo, *Traces of Bygone Biospheres*, 58.

75. Dietmar Offenhuber, *Autographic Design. The Matter of Data in a Self-Inscribing World.* (Cambridge, MA: MIT Press, 2023). Lukáš Likavčan and Paul Heinicker, "Planetary Diagrams: Towards an Autographic Theory of Climate Emergency," in *Photography Off the Scale: Technologies and Theories of the Mass Image*, ed. Tomáš Dvořák and Jussi Parikka (Edinburgh: Edinburgh University Press, 2021), 211–230; Susan Schuppli, *Material Witness: Media, Forensics, Evidence* (Cambridge, MA: MIT Press, 2020).

76. Vernadsky "Evolution of Species and Living Matter," 44. See also Vernadsky, *The Biosphere*, 60.

77. Vernadsky, *The Biosphere*, 51.

78. Vernadsky *The Biosphere*, 60.

79. Vernadsky, *The Biosphere*, 66–67.

80. Vernadsky, *The Biosphere*, 65.

81. Based on this notion, Vernadsky defined the time that it would take for a living form to occupy the whole Earth, if it did not have predators or limited resources (*The Biosphere*, 66). He calculated this time for some species, such as certain bacteria—36 hours—or termites—"a few years" (63).

82. Vernadsky, *The Biosphere*, 61.

83. Leslie, *Synthetic Worlds.*

84. Vernadsky, "Biosphere and the Noösphere," 9.

85. John Bellamy Foster, *Marx's Ecology: Materialism and Nature* (New York: Monthly Review Press, 2000), ix.

86. Leslie, *Synthetic Worlds*, 82.

87. Vaclav Smil, "Population Growth and Nitrogen: En exploration of a Critical Existential Link," *Population and Development Review* 17, no. 4 (1991): 569–601. Leslie, *Synthetic Worlds*, 184–188.

88. John H. Perkins, *Geopolitics and the Green Revolution: Wheat, Genes and the Cold War* (New York: Oxford University Press, 1997)

89. Paul Crutzen, "Geology of Mankind," *Nature* 415 (2002), 23. Anthropogenic nitrogenation of soils is currently so excessive that overfertilization has been considered as one of the main environmental concerns today. See Fred Pearce, "Can the World Find Solutions to the Nitrogen Pollution Crisis?," *Yale Environment E360*, February 6, 2018, https://e360.yale.edu/features/can-the-world-find-solutions-to-the-nitrogen-pollution-crisis.

90. Henning, *Photography*, 95–99.

91. Henning, *Photography*, 99.

92. Vernadsky *The Biosphere*, 61.

93. Paul N. Edwards, "Infrastructure and Modernity: Force, Time, and Social Organization in the History of Sociotechnical Systems," in *Modernity and Technology*, ed. Thomas J. Misa, Philip Brey, and Andrew Feenberg, (Cambridge, MA: MIT Press, 2003), 185–225, at 196.

94. Lapo, *Traces of Bygone Biospheres*, 103.

95. Lapo, *Traces of Bygone Biospheres*, 103.

96. A telling example of this is how life needs to be protected from the otherwise excessive radiation of the sun. Unfiltered light, such as that experienced in outer space, is a lethal agent: no form of life is able to survive the effects of certain ultraviolet components of its spectrum. On the earth, the ozone layer is responsible for this radiation not reaching the surface of the planet. Significantly, it was Vernadsky who proved, with geochemical data, the theory that the totality of the ozone in the atmosphere had a biotic origin; Aleksandr I. Oparin, *The Origin of Life on the Earth*, trans. Ann Synge (New York: Academic Press, 1957), 157. That is, in the biogeochemist's model, it is not only that living matter keeps suitable environmental conditions but it creates its own shielding jar as well, the "ozone screen"; Vernadsky, *The Biosphere*, 120.

97. Vernadsky in 1913, quoted in G. V. Dobrovol'skii, "Vladimir Ivanovich Vernadsky and Soil Science," *Herald of the Russian Academy of Sciences* 83 (2013): 184.

98. Veronica della Dora, *The Mantle of the Earth: Genealogies of a Geographical Metaphor* (Chicago: University of Chicago Press, 2021), 228.

99. The following quotes and references are from the Russian original, Dokuchaev, *К учению о зонах природы. Горизонтальные и вертикальные почвенные зоны* (Theory of natural zones: Horizontal and vertical soil zones) (St. Petersburg: Printing House of St. Petersburg City Administration, 1899), https://ru.wikisource.org/wiki/К_учению_о_зонах_природы. We would like to thank Anastasia Kubrak for the translations. See also Wladimir Dokuchaev, *Tchernozéme (Terre Noire) de la Russie d'Europe* (St. Petersbourg: Imprimerie Trenké & Fusnot, 1879).

100. Dokuchaev, *К учению о зонах природы*.

101. Among the geographical surfaces that Dokuchaev investigated and coded were the horizontal zones of soil, for example, in the Caucasus Mountains.

102. Dokuchaev, *К учению о зонах природы*.

103. See for example Vernadsky, "Biosphere and the Noösphere."

104. John Durham Peters, *The Marvelous Clouds: Toward a Philosophy of Elemental Media* (Chicago: University of Chicago Press, 2015), 47. See also Jussi Parikka, *A Geology of Media* (Minneapolis: University of Minnesota Press, 2015).

105. "Neither living organisms by themselves nor their environment abstracted from them are, Vernadsky argued, the specific objects of biogeochemistry. A biogeochemist is interested, first of all, in studying the cyclic processes of the exchange of chemical elements between living organisms and their environment." Georgy S. Levit, "Looking at Russian Ecology Through the Biosphere Theory," in *Ecology Revisited: Reflecting on Concepts, Advancing Science*, ed. Astrid Schwarz

and Kurt Jax (Dordrecht: Springer, 2011), 333–347, at 339. Vernadsky went so far as to claim that all geological periods include some form of life: "Throughout geological time, no azoic (i.e., devoid of life) geological periods have ever been observed"; Vernadsky, *The Biosphere*, 54.

106. Bruno Latour, *Facing Gaia: Eight Lectures on the New Climatic Regime*, trans. Catherine Porter (Cambridge: Polity, 2017), 111–145.

107. John Ruskin, "The Storm-Cloud of the Nineteenth Century," two lectures delivered at the London Institution, February 4 and 11, 1884; https://www.gutenberg.org/files/20204/20204-h/20204-h.htm.

108. On such a notion of parasitism, see Michel Serres, *The Parasite* (Minneapolis: University of Minnesota Press, 2007), 77–85.

CHAPTER 4

1. Andrey Tarkovsky, *Sculpting in Time: Reflections on the Cinema*, trans. Kitty Hunter-Blair (New York: Alfred A. Knopf, 1987), 132.

2. José Manuel Mouriño, *Andrei Tarkovski y El Espejo: Estudio de Un Sueño* (Madrid: Círculo de Bellas Artes, 2018), 16.

3. Graig Uhlin, "Plant-Thinking with Film: Reed, Branch, Flower," in *The Green Thread: Dialogues with the Vegetal World*, ed. Patrícia Vieira, Monica Gagliano, and John Ryan (Lanham, MD: Lexington, 2015), 215. Moreover, as Matthew Fuller and Olga Goriunova write, the architectural trope of the home is closely aligned with the bordering force of the forest that speaks to an ecological aesthetic found in Tarkovsky and beyond, relating to questions of territory, property, belonging, and migrancy; Matthew Fuller and Olga Goriunova, *Bleak Joys: Aesthetics of Ecology and Impossibility* (Minneapolis: University of Minnesota Press, 2019), 121–153.

4. Tarkovsky, *Sculpting in Time*, 138–150.

5. "The plant's body is all skin," writes Michael Marder in *Plant-Thinking: A Philosophy of Vegetal Life* (New York: Columbia University Press, 2013), 81. "It is for the sake of adhering as much as possible to the world that they develop a body that privileges surface to volume," continues Emanuele Coccia in *The Life of Plants: A Metaphysics of Mixture* (Medford, MA: Polity, 2018), 5.

6. Uhlin, "Plant-Thinking with Film," 211–212.

7. Giuliana Bruno, *Surface: Matters of Aesthetics, Materiality, and Media* (Chicago: University of Chicago Press, 2014), 35–51.

8. Using Bruno's words, see Bruno, *Surface*, 13–22.

9. See, for example, Gilles Deleuze on the cinematics of water: "a perception not tailored to solids, which no longer had the solid as object, as condition, as milieu"; Gilles Deleuze, *Cinema 1: The Movement Image*, trans. Hugh Tomlinson and Barbara Habberjam (London: Continuum, 1992), 80.

10. Nadia Bozak, *The Cinematic Footprint: Lights, Camera, Natural Resources* (New Brunswick, NJ: Rutgers University Press, 2012), 1–11.
11. Bozak, *Cinematic Footprint*, 54.
12. James C. Scott, *Seeing Like a State: How Certain Schemes to Improve the Human Condition Have Failed* (New Haven, CT: Yale University Press, 1999), 53–63.
13. Ross Exo Adams, "Landscapes of Post-History," in *Landscape and Agency: Critical Essays*, ed. Ed Wall and Tim Waterman (Abingdon: Routledge, 2018), 7–17.
14. The Spanish law defined this inner colonization: "an administrative activity, of a technical and legal nature, that transforms the agronomic and economic characteristics of specific extensions of land as well as the social organisations within them, creating rational units of agricultural production whose property is delivered to certain farmers, with the aim to ease the fulfilment of their individual and familiar needs, provide stability to the society as a whole and augment the production"; see Alejo Leal García, "La transformación del medio rural a través de la puesta en regadío y de la colonización," *Revista de Estudios Agrosociales* 66 (1969): 107–137, at 116.
15. See Antonio A. Tordesillas, "Referencias internacionales en los pueblos de colonización españoles," *Ciudades* 13 (2010): 183–200, at 185. Two important exceptions are important to mention, as their influence on Spanish inner colonization has been highlighted several times. The first is the large North American irrigation program the Columbia Basin Project, whose final works took place between 1948 and 1952. It had been designed in the 1920s, but it was only after the war that water started to arrive to the fields. The second is the agrarian reform carried out by the Italian Christian Democrats after the war, as a sister project of the Spanish inner colonization. See also Gina Bloodworth and James White, "The Columbia Basin Project: Seventy-Five Years Later," *Yearbook of the Association of Pacific Coast Geographers* 70, no. 1 (2008): 96–111.
16. This intensification of productivity has been linked by writer and activist Vandana Shiva to the notion of the Bio Nullius, an extension of the colonial Terra Nullius, where "the Earth is defined by her colonizers as dead matter, deemed unable to create, and farmers [are] deemed to have empty heads that cannot innovate"; see Vandana Shiva, *Biopiracy: The Plunder of Nature and Knowledge* (Boston, MA: South End, 1999), 33. The old farmers, together with their old and unimproved nature, become the parasite of a parasitic activity or, as Shiva puts it, a "biopiracy." Shiva describes this process as a colonization of the interior spaces: "The colonies have now been extended to the interior spaces . . . from microbes and plants to animals, including humans" (40).
17. Among these *bonifiche*, the *integrale* (1924–1950) was the most influential for the Spanish program. See Cristóbal Gómez Benito, "Una revisión y una reflexión sobre la política de colonización agraria en la España de Franco," *Historia del presente* 3 (2004): 65–86, at 75.
18. Robert L. Nelson, "From Manitoba to the Memel: Max Sering, Inner Colonization and the German East," *Social History* 35, no. 4 (2010): 445.

19. Nelson, "From Manitoba to the Memel," 444.
20. Nelson, "From Manitoba to the Memel," 447. "Through an inner colonial programme, German settlers were to send wagons east and Germanize the land. With this move, Germany's relationship with its eastern borderlands began to resemble the world of New Imperialism and the link to the specific North American version of that global imperialism was never far away: in a booklet produced in October 1886 to help civil servants better understand their role in the new programme, a helpful list of seven books was provided. concerned the land and history of Posen and West Prussia. The initial six books concerned the land and history of Posen and West Prussia. The seventh was entitled *Manitoba and the Western Territories*" (447).
21. Tiago Saraiva, *Fascist Pigs: Technoscientific Organisms and the History of Fascism* (Cambridge, MA: MIT Press, 2016).
22. Gómez Benito, "Una revisión y una reflexión," 66.
23. Gómez Benito, "Una revisión y una reflexión," 109.
24. As an activity developed over three weekends between April and June of 2017, one of the authors proposed digitizing and uploading with creative commons license the data relative to the INC. The proposal was accepted as part of a collaborative production workshop at Medialab-Prado in Madrid, where a first complete digital inventory of the settler towns, together with other infrastructures, was produced. Andrés Rodríguez Muñoz, Marco Rizzetto, Carmen M. Pellicer Balsalobre, Guillermo Cid, and David Prieto were the other members of the team behind this project. It was an open-data initiative that can be accessed at https://medialab-prado.github.io/poblados-colonizacion-colonias-penitenciarias/index.html.
25. Gómez Benito, "Una revisión y una reflexión," 72.
26. The INC was not in charge of the main water infrastructures, such as dams and reservoirs. This was the task of the Ministerio de Obras Públicas y Urbanismo (ministry of public infrastructure, MOPU).
27. Gómez Benito, "Una revisión y una reflexión," 84.
28. It is necessary to clarify that during the existence of Spanish Inner Colonization, its methodology and scope did not remain constant. Three periods have been distinguished, of approximately ten years each (Eduardo Delgado, *Imagen y memoria: Fondos del archivo fotográfico del Instituto Nacional de Colonización, 1939–1973* [Madrid: Ministerio de Agricultura, Alimentación y Medio Ambiente Centro de Publicaciones, 2013], 79–80). In brief, the first decade was characterized by an autarchy model confident of private initiative: landowners, supported by public resources, would lead the transformation of the land and settle the peasants. The model proved to be a total failure (Tordesillas, "Referencias internacionales," 191) and, beginning in the 1950s, a new law was proposed, in which the state was much more engaged. This second period turned out to be the golden decade of colonization (Delgado, *Imagen y memoria*, 80): big plans, such as the Plan Jaén and Plan Badajoz, were implemented and were coupled

with industrialization programs. In 1962, however, these plans were criticized by a World Bank report as they involved large amounts of national debt. As a consequence, new economic criteria started to be used to evaluate the program, which resulted in the estrangement of the whole project (see Alfredo Villanueva Paredes and Jesús Leal Maldonado, eds., *Historia y evolucion de la colonizacion agraria en España III*, vol. 3, *La planificación del regadio y los pueblos de colonización* (Madrid: Ministerio de Agricultura, Pesca y Alimentación, 1990). Therefore, during the last decade, the INC became a passive technical manager, a mediator, in fact, that gradually facilitated the introduction of big food and agriculture companies in a progressively liberalized economy. See Mario Gaviria, José Manuel Naredo and Juan Serna, *Extremadura saqueada: Recursos naturales y autonomía regional* (Barcelona: Ruedo Ibérico, 1978), 435.

29. Gómez Benito, "Una revisión y una reflexión," 83.

30. Gaviria, Naredo, and Serna, *Extremadura saqueada*, 262.

31. On the evolution of the program, see Carlos Giménez and Luciano Sánchez, eds., *Historia y evolucion de la colonizacion agraria en España IV: Unidad y diversidad en la colonización agraria. Perspectiva comparada del desarrollo de las zonas regables*, vol. 4 (Madrid: Ministerio de Agricultura, Pesca y Alimentación, 1994). On its extractivist and repressive nature, see Gaviria, Naredo, and Serna, *Extremadura saqueada*. On the use of prisoners as slave labor, see José Luis Gutiérrez, Ángel del Río Sánchez, Gonzalo Acosta Bono, and Lola Martínez Macías, eds., *El canal de los presos, 1940–1962. Trabajos forzados: De la represión política a la explotación económica* (Barcelona: Editorial Crítica, 2004). On the relationship with previous colonization programs, see Carlos Barciela López, "La contrarreforma agraria y la política de colonización del primer franquismo, 1936–1959," in *Reformas y políticas agrarias en la historia de España: De la ilustración al primer franquismo*, ed. Ángel García and Jesús Sanz (Madrid: Centro de Publicaciones Agrarias, Pesqueras y Alimentarias, 1996), 351–398. On similar international developments, see Tordesillas, "Referencias internacionales." On its evolution and impact decades later, see Gómez Benito, "Una revisión y una reflexión," 74–84.

32. Delgado, *Imagen y memoria*.

33. Francisco de los Ríos Romero, "Colonización de las bardenas, cinco villas, somontano y Monegros," *Cuadernos de Aragón* 1 (1966): 181–230, at 229–230. The quote from 1966 coincides with the fundamentalist turn concerning land that in Israel—under the Zionist maxim of "making the desert bloom"—and in other countries linked agricultural production to a divine dictate; see Eyal Weizman and Fazal Sheikh, *The Conflict Shoreline: Colonialism as Climate Change in the Negev Desert* (Göttingen: Steidl, 2015), 8.

34. Lino Camprubí, *Engineers and the Making of the Francoist Regime* (Cambridge, MA: MIT Press, 2014).

35. Brenna Bhandar, *Colonial Lives of Property: Law, Land, and Racial Regimes of Ownership* (Durham, NC: Duke University Press, 2018), 77–114. Nelson, "From Manitoba to the Memel," 439–440.

36. The Ley de Colonización de Grandes Zonas (Law for the colonization of large areas) was passed in 1949. Its main aspects are summarized in Jesús González Pérez, "La colonización en zonas regables: La ley de 21 de abril de 1949," *Revista de estudios políticos* 48 (1949): 154–170, at 154–155.
37. On the relation between (landscape) painting and colonial land practices such as plantations, see Jill Casid, *Sowing Empire: Landscape and Colonization*. (Minneapolis: University of Minnesota Press, 2005), 30–93.
38. José Guarc Pérez, "El Instituto Nacional de Colonización y la transformación de Bardenas-Ejea," in *Colonos: Territorio y estado. Los pueblos del Agua de Bardenas*, ed. Alberto Sabio (Zaragoza: Instituto Fernando el Católico, 2010), 81–142, at 104.
39. Miguel Ángel Baldellou, "Prólogo," in Delgado, *Imagen y memoria*, 13–23, at 19.
40. Gaviria, Naredo, and Serna, *Extremadura saqueada*, 356–359.
41. Bernhard Siegert, *Cultural Techniques: Grids, Filters, Doors, and Other Articulations of the Real*, trans. Geoffrey Winthrop-Young (New York: Fordham University Press, 2015); Jussi Parikka, *Operational Images: From the Visual to the Invisual* (Minneapolis: University of Minnesota Press, 2023). See especially chapter 1 for a discussion of grids as operative ontologies.
42. Karsten Jacobsen, Michael Cramer, Richard Ladstädter, Camillo Ressl, and Volker Spreckels, "DGPF-Project: Evaluation of Digital Photogrammetric Camera Systems—Geometric Performance," *Photogrammetrie—Fernerkundung—Geoinformation* 2010, no. 2 (2010): 83–97, at 84.
43. Paul Saint-Amour, "Applied Modernism: Military and Civilian Uses of the Aerial Photomosaic," *Theory, Culture & Society* 28, no. 7–8 (2011): 241–269, at 243.
44. Lorenzo Martín-Retortillo Baquer, "Trayectoria y significación de las confederaciones hidrográficas," *Revista de Administración Pública* 25 (1958): 85–126, at 105.
45. Felipe Fernández García, "Las primeras aplicaciones civiles de la fotografía aérea en España: El catastro y las confederaciones hidrográficas," *Ería* 46 (1998): 117–130. One of the partners and initial president of the company CEFTA was Julio Ruiz de Alda (Fernández, "Las primeras aplicaciones," 222), one of the founders of the fascist movement in Spain, the Falange Española. He was a pioneer of aviation in Spain and a personal friend of the then dictator José Antonio Primo de Rivera and very close to his fascist government; see Barciela, "La contrarreforma agraria y la política de colonización," 355.
46. Francisco Quirós Linares and Felipe Fernández García, "El vuelo fotográfico de la 'Serie A,'" *Ería* 43 (1997): 190–198, at 190.
47. On the uses in Spain of the *Casey Jones* aerial photographs, see Juan Antonio Pérez, Francisco Manuel Bascón, and María Cristina Charro, "Photogrammetric Usage of 1956–57 USAF Aerial Photography of Spain," *Photogrammetric Record* 29, no. 145 (2014): 108–124. On its relation to a potential conflict with the USSR, see Quirós and González, "El vuelo fotográfico de la 'Serie A.'"
48. Pérez, Bascón, and Charro, "Photogrammetric Usage of 1956–57 USAF Aerial Photography of Spain," 22.

49. Fernández, "Las primeras aplicaciones," 118.

50. Guarc Pérez, "El Instituto Nacional de Colonización y la transformación de Bardenas-Ejea," 118. Figure 4.2 illustrates the transformation of the landscape in this particular zone.

51. On the excluded third, see Michel Serres, *The Parasite*, trans. Lawrence R. Schehr (Baltimore, MD: Johns Hopkins University Press, 1982).

52. See for example Alexander Etkind, *Internal Colonization: Russia's Imperial Experience* (Cambridge: Polity, 2011). For more recent discussion in light of the war in Ukraine, see Anna Engelhardt, "War by Any Other Name: Patterns of Russian Colonialism," *The Funambulist* 42 (July–August 2022): 10–13.

53. On the operationalization of the "hinterlands" of urbanism and production of industrial zones of extraction, see Neil Brenner and Nikos Katsikis, "Hinterlands of the Capitalocene," *Architectural Design* 90 (2020): 22–31.

54. In the context of art and design, for example, Benedikt Groß's project *Avena + Test Bed* (2013) employs such methods for large-scale landscape printing as a form of data-driven visual print media; https://benedikt-gross.de/projects/avena-test-bed-agricultural-printing-and-altered-landscapes/.

55. Keller Easterling, *Extrastatecraft: The Power of Infrastructure Space* (London: Verso, 2014), 19.

56. Easterling, *Extrastatecraft*, 22.

57. The new naming system was drawn up in the Plan General de Obras Públicas (1940). It referenced each road with a number that depended on (1) its angular location in relation to the six main radial motorways and (2) its distance to the Puerta del Sol square in Madrid. This nomenclature protocol is still in operation today.

58. Other geographical references were also used, such as the course of rivers or borderlines between provinces.

59. Several towns created during these plans by the Francoist dictatorship received names linked to the colonization of America, as a celebration of an infamous "glorious" national past; see Delgado, *Imagen y memoria*, 21.

60. Penny Harvey and Hannah Knox, *Roads: An Anthropology of Infrastructure and Expertise* (Ithaca, NY: Cornell University Press, 2015).

61. Nigel Thrift, "Movement-Space: The Changing Domain of Thinking Resulting from the Development of New Kinds of Spatial Awareness," *Economy and Society* 33, no. 4 (2004): 582–604, at 597.

62. Celia Lury, Luciana Parisi, and Tiziana Terranova, "Introduction: The Becoming Topological of Culture," *Theory, Culture & Society* 29, nos. 4–5 (2012): 3–35, at 5.

63. The design decision of positioning settlement towns by the existing road system has been discussed by Scott as the control and monitoring disposition of authoritarian states; see Scott, *Seeing Like a State*, 237. In this sense, the INC program has been criticized for its pervasive practices of surveillance on the settlers; see Gaviria, Naredo, and Serna, *Extremadura saqueada*, 356.

64. John Durham Peters, *The Marvelous Clouds: Toward a Philosophy of Elemental Media* (Chicago: University of Chicago Press, 2015), 37.

65. Ned Rossiter, *Software, Infrastructure, Labor: A Media Theory of Logistical Nightmares* (New York: Routledge, 2016), 6.

66. Serres, *The Parasite*, 179.

67. Tordesillas, "Referencias internacionales," 192.

68. José María Alagón Laste, "Los pueblos de colonización del Plan de Riegos del Alto Aragón y su emplazamiento en el territorio," *Scripta Nova* 19 (2015): 500–526, at 507.

69. Alagón Laste, "Los pueblos de colonización del Plan de Riegos del Alto Aragón," 508.

70. Lisa Parks, *Cultures in Orbit: Satellites and the Televisual* (Durham, NC: Duke University Press, 2005); Parks, "Signals and Oil: Satellite Footprints and Post-Communist Territories in Central Asia," *European Journal of Cultural Studies* 12, no. 2 (2009): 137–156.

71. The "cart" in the cart-module was the main transportation means in rural Spain at the time, a cart pulled by a donkey, horse, or farmer. As explained, the plans clearly took this into account. It was so central to the model that, as has been repeatedly argued, it was also its most salient failure, since the design became immediately obsolete with the widespread introduction of mechanized vehicles, following the end of the international blockade of Francoist Spain. See Alagón Laste, "Los pueblos de colonización del Plan de Riegos del Alto Aragón," 508; and Tordesillas, "Referencias internacionales," 199.

72. Cornelia Vismann, "Cultural Techniques and Sovereignty," *Theory, Culture & Society* 30, no. 6 (2013): 83–93, at 92.

73. Gaviria, Naredo and Serna, *Extremadura saqueada*, 356.

74. To keep the big landowners happy, the state assigned the worst lands to the settlers. The rest of the plots—still private property—were connected to the irrigation system. This way, the owners multiplied their benefits, despite the lands having been expropriated. Gaviria, Naredo and Serna, *Extremadura saqueada*, 262.

75. Stefan Ouma, *Farming as Financial Asset: Global Finance and the Making of Institutional Landscapes* (Newcastle: Agenda, 2020). See also Kelly Bronson and Phoebe Sengers, "Big Tech Meets Big Ag: Diversifying Epistemologies of Data and Power," *Science as Culture* 31, no. 1 (2022): 15–28.

76. Tarkovsky, *Sculpting in Time*, 138–150.

77. Tarkovsky, *Sculpting in Time*, 120. This quote is also the starting point of Gil-Fournier's installation *The Quivering of the Reed*, 2019, http://abelardogfournier.org/works/quivering.html.

78. Barbara Novak, *Nature and Culture American Landscape and Painting, 1825–1875*, 3rd ed. (Oxford: Oxford University Press, 2007), 155. On Muybridge, Yosemite, and landscapes, see also Bozak, *Cinematic Footprint*, 93–95.

79. This tension is one of the issues we addressed in our video essay *Seed, Image, Ground* (2020). The video discusses the technique known as seed bombing, the dropping of biodegradable containers filled with seeds and mineral nutrients from aerial vehicles—drones, helicopters, or aircrafts. It is a technique used in forestry and environmental restoration, and its most outstanding characteristic, according to its promoters, is the speed with which it can repopulate when compared to other techniques. Based on communication and advertisement videos of these technologies, and through the image comparison technique conceptualized by Harun Farocki as soft montage, the visual essay outlines a history in which the image has been used not only to catalog and classify plants and species but also to accelerate their growth. In particular, the essay shows how an understanding of the landscape is linked to methods of analysis based on aerial photography. From early twentieth-century laboratory practices to the cameras, sensing practices, and computational analytics involved now, seed bombing has shifted from a "guerrilla" agricultural practice to part of the broader aerial management exemplified in precision farming. The term seed bombing alludes to the military history of the aerial view, yet it also expands into the agricultural practices as a history of accelerating or annihilating growth. Thus, it is also a visual montage version of the argument presented in this chapter: techniques of management of agricultural lands are inherently linked with operational techniques of visual definition, observation, and control.
80. On this question in history of biology, see Robert E. Kohler, *Landscapes and Labscapes: Exploring the Lab-Field Border in Biology* (Chicago: University of Chicago Press, 2002).
81. Commenting on an advertisement in the December 1915 issue of the *Tractor Farming* magazine, the historian of architecture and technology Siegfried Giedion wrote: "to rouse the farmer's imagination, a tractor and an airplane are shown side by side with the comment: 'This butterfly and this ant are sisters under the skin'", Siegfried Giedion, *Mechanization Takes Command: A Contribution to Anonymous History* (New York: Oxford University Press, 1970), 162.
82. For an overview, see for example Abdul Hafeez, Mohammed Aslam Husain, S. P. Singh, Anurag Chauhan, Mohammed Tauseef Khan, Navneet Kumar, Abhishek Chauhan, and S. K. Soni, "Implementation of Drone Technology for Farm Monitoring & Pesticide Spraying: A Review," *Information Processing in Agriculture* 10, no. 2 (2023): 192–203.

CHAPTER 5

1. Joshua DiCaglio, *Scale Theory: A Nondisciplinary Inquiry* (Minneapolis: University of Minnesota Press 2021), 7. See also Zachary Horton, *The Cosmic Zoom: Scale, Knowledge, and Mediation* (Chicago: University of Chicago Press, 2021). For a cautionary take on the concept of scale, see Anna Tsing, "On Nonscalability: The Living World Is Not Amenable to Precision-Nested Scales," *Common Knowledge* 18 (2012): 505–524. See also Dipesh Chakrabarty, "Afterword: On Scale and

Deep History in the Anthropocene," in *Narratives of Scale in the Anthropocene: Imagining Human Responsibility in an Age of Scalar Complexity*, ed. Gabriele Dürbeck and Philip Hüpkes (London: Routledge, 2021), 225–232. Jussi Parikka, *There is Plenty of Room in the Simulation* (Ljubljana: Aksioma, 2023), https://aksioma.org/there-is-plenty-of-room-in-the-simulation.

2. See, for example, Laura Kurgan, *Close up at a Distance: Mapping, Technology, and Politics* (New York: Zone, 2013); Pamela E. Mach, *Viewing the Earth: The Social Construction of the Landsat Satellite System* (Cambridge, MA: MIT Press, 1990); Mariel Borowitz, *Open Space: The Global Effort for Open Access to Environmental Satellite Data* (Cambridge, MA: MIT Press, 2017).
3. John May, "Logic of the Managerial Surface," *PRAXIS* 13 (2012): 116–124; Jennifer Gabrys, *Program Earth: Environmental Sensing Technology and the Making of a Computational Planet* (Minneapolis: University of Minnesota Press, 2016).
4. In relation to this, see Michel Foucault on territory and other terms of relevance: "Territory is no doubt a geographical notion, but it's first of all a juridico-political one: the area controlled by a certain kind of power. Field is an economico-juridical notion. Displacement: what displaces itself is an army, a squadron, a population. Domain [*domaine*] is a juridico-political notion. Soil is a historico-geological notion. Region is a fiscal, administrative, military notion. Horizon is a pictorial, but also a strategic notion"; Michel Foucault, "Questions of Geography," in *Space, Knowledge, and Power: Foucault and Geography*, ed. Jeremy W. Crampton and Stuart Elden, trans. Colin Gordon (Aldershot: Ashgate, 2007), 176.
5. Bo Zhao, Shaozeng Zhang, Chunxue Xu, Yifan Sun, and Chengbin Deng, "Deep Fake Geography? When Geospatial Data Encounter Artificial Intelligence," *Cartography and Geographic Information Science* 48, no. 4 (2021): 338–352.
6. Cindy Lin, "How Forests Became Data: The Remaking of Ground Truth in Indonesia," in *The Nature of Data: Infrastructures, Environments, Politics*, ed. Jenny Goldstein and Eric Host (Lincoln: University of Nebraska Press, 2022), 285–302.
7. See Gillian Rose, *Feminism and Geography: The Limits of Geographical Knowledge* (Minneapolis: University of Minnesota Press, 1993); Irit Rogoff, *Terra Infirma: Geography's Visual Culture* (London: Routledge, 2000); Nigel Thrift, *Non-Representational Theory: Space, Politics, Affect* (London: Routledge, 2008).
8. Bernhard Siegert, "The Map Is the Territory," *Radical Philosophy* 169, no. 5 (2011): 13–16. See also Shannon Mattern, *Code and Clay, Data and Dirt: Five Thousand Years of Urban Media* (Minneapolis: University of Minnesota Press, 2017).
9. Liam Young, "An Atlas of Fiducial Landscapes: Touring the Architectures of Machine Vision" *Log* 36 (Winter 2016): 125–134.
10. See John Pickles, ed., *Ground Truth: The Social Implications of Geographic Information Systems* (New York: Guilford, 1994).
11. Jean-Luc Nancy, *The Ground of the Image*, trans. Jeff Fort (New York: Fordham University Press, 2005). For a related discussion on the troubling of this figure/ground distinction leading to the elaboration of legal infrastructure figures out

of invisible underground formations, see Andrea Ballestero, "Underground as Infrastructure? Figure/Ground Reversals and Dissolution in Sardinal," in *Environment, Infrastructure and Life in the Anthropocene*, ed. Kregg Hetherington (Durham, NC: Duke University Press, 2019), 17–44.

12. Nancy, *Ground of the Image*, 13.
13. Harun Farocki, "Phantom Images," trans. Brian Poole, *Public* 29 (2004): 12–22; Jussi Parikka, *Operational Images: From the Visual to the Invisual* (Minneapolis: University of Minnesota Press, 2023).
14. Machine vision and machine learning do not necessarily lead to images but to all sorts of elevation data, point clouds, models, as well as other higher-dimensional entities; so the technique of flattening takes place not as a cut in the projective space of representation but as a recursive operation, even at the level of hardware, as for instance in the graphical parallel computation performed by graphics processing units. See Jacob Gaboury, *Image Objects: An Archaeology of Computer Graphics* (Cambridge, MA: MIT Press, 2021).
15. Seán Cubitt, "Mass Image, Anthropocene Image, Image Commons," in *Photography Off the Scale: Technologies and Theories of the Mass Image*, ed. Tomáš Dvořák and Jussi Parikka (Edinburgh: University of Edinburgh Press, 2021), 25–40.
16. Caren Kaplan, *Aerial Aftermaths: Wartime from Above* (Durham, NC: Duke University Press, 2018), 34.
17. See, for example, the case of Landsat, 1970s, discussed in Mack, *Viewing the Earth.*
18. Indeed, as is convincingly shown in many studies and contexts, observations are always theory-dependent and part of a more detailed back and forth movement of comparison and synthesis in contexts of the materiality of epistemic practices. See, for instance, Karin Knorr-Cetina and Michael Mulkay, *Science Observed: Perspectives on the Social Study of Science* (London: SAGE, 1983); Ian Hacking, *Representing and Intervening: Introductory Topics in the Philosophy of Natural Science* (Cambridge: Cambridge University Press, 1983); Bruno Latour and Steve Woolgar, *Laboratory Life: The Construction of Scientific Facts* (Princeton, NJ: Princeton University Press, 1986).
19. Thomas M. Lillesand, Ralph W. Kiefer, and Jonathan W. Chipman, *Remote Sensing and Image Interpretation*, 7th ed. (Hoboken, NJ: Wiley, 2015), 39.
20. Roger Hoffer, "The Importance of 'Ground Truth,'" *Data in Remote Sensing* 120371 of LARS Print (1972): 1–13, at 3.
21. Mack, *Viewing the Earth*, 49.
22. Mack, *Viewing the Earth.*
23. EROS CalVal Center of Excellence, "Test Sites Catalog," *U.S. Geological Survey*, accessed June 9, 2022. https://calval.cr.usgs.gov/apps/test_sites_catalog. The following quotations are from the catalog.
24. A Google Ngram search showing the use of several expressions for the same concept displays the growing popularity of the term ground truth since the 1960s. For more on n-grams, see Jean-Baptiste Michel, Yuan Kui Shen, Aviva

Presser Aiden, Adrian Veres, Matthew K. Gray, Joseph P. Pickett, Dale Hoiberg, et al., "Quantitative Analysis of Culture Using Millions of Digitized Books," *Science* 331, no. 6014 (2011): 176.

25. Parikka, *Operational Images*, 118–126.
26. Denis Cosgrove and William L. Fox, *Photography and Flight* (London: Reaktion, 2010).
27. Paul K. Saint-Amour, "Modernist Reconnaissance," *Modernism/Modernity* 10, no. 2 (2003): 349–380; Saint-Amour, "Applied Modernism: Military and Civilian Uses of the Aerial Photomosaic," *Theory, Culture & Society* 28, no. 7–8 (2012): 241–269; Saint-Amour, "Photomosaics: Mapping the Front, Mapping the City," in *From Above: War, Violence, and Verticality*, ed. Peter Adey, Mark Whitehead, and Alison Williams (Oxford: Oxford University Press, 2014), 119–142.
28. Saint-Amour, "Modernist Reconnaissance," 356
29. Saint-Amour, "Modernist Reconnaissance," 354.
30. Saint-Amour, "Modernist Reconnaissance," 360–361.
31. Saint-Amour, "Modernist Reconnaissance," 357.
32. Saint-Amour, "Modernist Reconnaissance," 358.
33. Saint-Amour, "Modernist Reconnaissance," 358.
34. Saint-Amour, "Photomosaics."
35. Gabrys, *Program Earth*, 71.
36. John Pickles, ed., *Ground Truth: The Social Implications of Geographic Information Systems* (New York: Guilford, 1995).
37. Howard Veregin, "Computer Innovation and Adoption in Geography: A Critique of Conventional Technological Models," in Pickles, *Ground Truth*, 100–101.
38. Paul N. Edwards, *A Vast Machine: Computer Models, Climate Data, and the Politics of Global Warming* (Cambridge, MA: MIT Press, 2010), xiii.
39. Carlo Ginzburg, *Clues, Myths, and the Historical Method*, trans. John Tedeschi and Anne C. Tedeschi (Baltimore, MD: Johns Hopkins University Press, 2013).
40. Eyal Weizman, *Forensic Architecture: Violence at the Threshold of Detectability* (New York: Zone, 2017); Susan Schuppli, *Material Witness: Media, Forensics, Evidence* (Cambridge, MA: MIT Press, 2020).
41. John K. St. Joseph, "Air Photography and Archaeology," *Geographical Journal* 105, no. 1/2 (1945): 47–59.
42. Margaret Cox, Ambika Flavel, and Ian Hanson, *The Scientific Investigation of Mass Graves: Towards Protocols and Standard Operating Procedures* (Cambridge: Cambridge University Press, 2008).
43. Umberto Lombardo, José Iriarte, Lautaro Hilbert, Javier Ruiz-Pérez, José M. Capriles, and Heinz Veit, "Early Holocene Crop Cultivation and Landscape Modification in Amazonia," *Nature* 581, no. 7807 (2020): 190–193.
44. Schuppli, *Material Witness*, 3.

45. Susan Schuppli, "Impure Matter: A Forensics of WTC Dust," in *Savage Objects*, ed. Godofredo Perreira (Lisbon: Imprensa Nacional Casa da Moeda, 2012), 119–140.

46. "Where on Earth? Quizzes," California Institute of Technology: Jet Propulsion Laboratory, MISR (Multi-Angle Imaging Spectroradiometer), accessed June 22, 2020, https://misr.jpl.nasa.gov/search/ ? keyword=quiz.

47. Tobias Weyand, Ilya Kostrikov, and James Philbin, "PlaNet—Photo Geolocation with Convolutional Neural Networks," in *Computer Vision—ECCV 2016*, ed. Bastian Leibe, Jiri Matas, Nicu Sebe, and Max Welling, vol. 8 (New York: Springer, 2016), 37–55.

48. Adrian Mackenzie, *Machine Learners: Archaeology of a Data Practice* (Cambridge, MA: MIT Press, 2017).

49. Adrian Mackenzie and Anna Munster, "Platform Seeing: Image Ensembles and Their Invisualities," *Theory, Culture & Society* 36, no. 5 (2019): 18. See also Parikka, *Operational Images*.

50. See Ryan Bishop, "Project 'Transparent Earth' and the Autoscopy of Aerial Targeting: The Visual Geopolitics of the Underground," *Theory, Culture & Society* 28, no. 7–8 (2011): 270–286. On the notion of data ensemble, see Ingrid Hoelzl and Rémi Marie, "Google Street View: Navigating the Operative Image," *Visual Studies* 29 (2014): 261–271. On image ensembles, see Mackenzie and Munster, "Platform Seeing."

51. Alexis C. Madrigal, "How Google Builds Its Maps—And What It Means for the Future of Everything," *The Atlantic*, September 6, 2012; https://www.theatlantic.com/technology/archive/2012/09/how-google-builds-its-maps-and-what-it-means-for-the-future-of-everything/261913/.

52. Stefan Strauß, "From Big Data to Deep Learning: A Leap Towards Strong AI or 'Intelligentia Obscura'?" *Big Data and Cognitive Computing* 2, no. 16 (July 17, 2018): 11.

53. Andrew Lookingbill and Michael Weiss-Malik, "Google I/O 2013—Project Ground Truth: Accurate Maps Via Algorithms and Elbow Grease," Google for Developers, May 16, 2013, YouTube video, 40:20, https://www.youtube.com/watch?v=FsbLEtSouls.

54. Kate Crawford and Vladan Joler, "Anatomy of an AI System: The Amazon Echo as an Anatomical Map of Human Labor, Data and Planetary Resources," *AI Now Institute and Share Lab*, September 7, 2018, https://anatomyof.ai/; Maya Indira Ganesh, "Intelligence Work," *A is for Another: A Dictionary of AI*, accessed 8 April, 2023 https://aisforanother.net/pages/article16.html; Matteo Pasquinelli and Vladan Joler, "The Nooscope Manifested: AI as Instrument of Knowledge Extractivism," KIM HfG Karlsruhe and Share Lab, May 1, 2020, accessed 14 Nov, 2022, https://nooscope.ai/. See also Lin, "How Forests Became Data."

55. Bishop, "Project 'Transparent Earth,'" 276.

56. Walter G. Eppler and R. D. Merrill. "Relating Remote Sensor Signals to Ground-Truth Information," *Proceedings of the IEEE* 57, no. 4 (1969): 665–675, at 665.

57. Robert L. Grossman and William E. Marlatt, "A Method of Showing What a Radiometer 'Sees' during an Aircraft Survey," *Proceedings of the Fourth Symposium on Remote Sensing of Environment*, 1966, Ann Arbor, University of Michigan, 571–574.

58. Not by accident, these techniques of overlapping images and data in the same frame were published simultaneously with Ian L. McHarg's seminal book, *Design with Nature* (Garden City, NY: American Museum of Natural History, 1969), which proposed the layer-cake model, acknowledged as a forerunner of GIS. The layer-cake model diagrammed representations of different components of a natural environment as different layers. Geology, soils, vegetation, and climate were portrayed as layers superimposed on each other in order to emphasize the viewing of the composite and the relations between the components. See Frederick Steiner and Billy Fleming, "*Design with Nature* at 50: Its Enduring Significance to Socio-Ecological Practice and Research in the Twenty-First Century," *Socio-Ecological Practice Research* 1, no. 3 (October 1, 2019): 173–177.

59. On invisual, see Mackenzie and Munster, "Platform Seeing."

60. Lin, "How Forests Became Data," 291.

61. Paul Virilio, *The Vision Machine*, trans. Julie Rose (Bloomington: Indiana University Press, 1994), 13.

62. Padhraic Smyth, Usama Fayyad, Michael Burl, Pietro Perona, and Pierre Baldi, "Inferring Ground Truth from Subjective Labelling of Venus Images," *Proceedings of the 7th International Conference on Neural Information Processing Systems* (Cambridge, MA: MIT Press, 1994), 1085–1092, at 1086.

63. Smyth et al., "Inferring Ground Truth from Subjective Labelling of Venus Images."

64. Bernhard Siegert, *Cultural Techniques: Grids, Filters, Doors, and Other Articulations of the Real*, trans. Geoffrey Winthrop-Young (New York: Fordham University Press, 2015).

65. J. F. Bell III, S. W. Squyres, Kenneth E. Herkenhoff, J. N. Maki, H. M. Arneson, D. Brown, S. A. Collins, et al. "Mars Exploration Rover Athena Panoramic Camera (Pancam) Investigation," *Journal of Geophysical Research* 108, no. E12 (2003): 10–11.

66. Nadia Bozak, *The Cinematic Footprint: Lights, Camera, Natural Resources* (New Brunswick, NJ: Rutgers University Press, 2011), 34.

67. On the question of calibration both as a technical and artistic notion, see Geocinema's work *Framing Territories* (2019), which focuses on remote sensing and science infrastructures of the Digital Belt and Road, and Hito Steyerl's *How Not to Be Seen: A Fucking Didactic Educational .MOV File* (2013), https://www.artforum.com/video/hito-steyerl-how-not-to-be-seen-a-fucking-didactic-educational-mov-file-2013-51651.

68. Lukáš Likavčan, *Introduction to Comparative Planetology* (Moscow: Strelka, 2019).

69. Zhao et al., "Deep Fake Geography?," 338.

70. Zhao et al., "Deep Fake Geography?," 338. The authors explain the typical GAN-created synthetic media process: "The GANs generate two networks—a 'generator' and a 'discriminator'—and enable them to contest with one another through a multiple-epoch training process. In the training process, the generator creates a latent space of candidate datasets, and then the discriminator evaluates whether the candidate datasets are qualified by satisfying an evolving statistical characteristics criterion. The candidate data from the generator, after several training epochs of tuning, can reach an acceptable similarity to the required statistical characteristics. . . . Similarly, if we use a GAN to simulate geospatial data, the GAN's generator will create candidates of geospatial data and ask the discriminator whether the candidates meet the characteristics of a typical geospatial data. Here, the geospatial data can be as simple as a point, polyline or polygon, or relative complex data like satellite images, or even 3D point clouds. After several epochs' training, the candidates could eventually meet the criteria of qualified geospatial data. At this stage, the candidates, recognized as seemingly authentic geospatial data, embody a new mode of fake geography" (341).

71. See "Fake Geography," https://chunxxu.github.io/fakegeo/index.html.

72. Chunxue Xu and Bo Zhao, "Satellite Image Spoofing: Creating Remote Sensing Dataset with Generative Adversarial Networks," *10th International Conference on Geographic Information Science, GIScience*, ed. Stephan Winter, Amy Griffin, and Monika Sester (Wadern: Leibniz-Zentrum fur Informatik, 2018), 67:1–67:6.

73. Bo Zhao et al., "Deep Fake Geography?"

74. Weili Shi, "Terra Mars: When Earth Shines on Mars through AI's Imagination," *Leonardo* 52, no. 4 (2019): 357–363.

75. Casey Handmer, "Terraformed Mars," Twitter bot, 2018, accessed June 22, 2020, https://twitter.com/terraformedmars.

76. Bozak, *Cinematic Footprint*, 97.

77. See chapter 4 for the discussion on *Seed, Image, Ground* (2020). On soft montage, see Harun Farocki, "Cross Influence/Soft Montage," in *Harun Farocki: Against What? Against Whom?* Ed. Antje Ehmann and Kodwo Eshun (London: Koenig, 2009), 64–79; Volker Pantenburg, "Working Images: Harun Farocki and the Operational Image," in *Image Operations: Visual Media and Political Conflict*, ed. Jens Eder and Charlotte Klonk (Manchester: Manchester University Press, 2017), 49–62; Nora Alter, "Two or Three Things I Know about Harun Farocki," *October* 151 (Winter 2015): 151–158.

78. Nancy, *Ground of the Image*, 13.

79. Cubitt, "Mass Image."

80. Mackenzie, *Machine Learners*, 53.

81. Mackenzie and Munster, "Platform Seeing." On visual culture and data, see also Parikka, *Operational Images*.

82. John Pickles, *A History of Spaces: Cartographic Reason. Mapping and the Geo-Coded World* (New York: Routledge, 2004), 159.

83. Irmgard Emmelhainz, "Conditions of Visuality under the Anthropocene and Images of the Anthropocene to Come," *e-flux* 63 (March 2015), https://www.e-flux.com/journal/63/60882/conditions-of-visuality-under-the-anthropocene-and-images-of-the-anthropocene-to-come/.

CHAPTER 6

1. Robert E. Kohler, *Landscapes and Labscapes: Exploring the Lab-Field Border in Biology* (Chicago: University of Chicago Press, 2002).
2. Alfred G. Tansley, "The Use and Abuse of Vegetational Concepts and Terms," *Ecology* 16, no. 3 (1935): 284–307, at 285.
3. Robert Gerard Pietrusko, "Ground Cover," *LA*+ Geo Issue (Fall 2020): 12–19; John May, "Logic of the Managerial Surface," *PRAXIS* 13 (2012), 116–124.
4. Suzanne Simard, *Finding the Mother Tree: Discovering the Wisdom of the Forest* (New York: Alfred A. Knopf, 2021), 3.
5. After his work on plant succession and plant indicators, Clements did work on the importance of the different densities of root connectors of plants in the prairies. His doctoral student John E. Weaver performed a systematic excavation of the roots of thousands of plants to confirm his ideas on the root networks that sustained the plant's ability to survive the year-long periods of abnormal weather. See Ronald C. Tobey, *Saving the Prairies: The Life Cycle of the Founding School of American Plant Ecology, 1895–1955* (Berkeley: University of California Press, 1981), 191–193.
6. Thomas Kirchoff, "The Myth of Frederic Clements's Mutualistic Organicism, or: On the Necessity to Distinguish Different Concepts of Organicism," *History and Philosophy of the Life Sciences* 42 (2020): 1–27.
7. Pietrusko, "Ground Cover."
8. Stefano Mancuso and Alessandra Viola, *Brilliant Green: The Surprising History and Science of Plant Intelligence*, trans. Joan Benham (Washington, DC: Island, 2015), 34.
9. Michael Marder, *Plant-Thinking: A Philosophy of Vegetal Life* (New York: Columbia University Press, 2013).
10. Ursula K. LeGuin, "Vaster than Empires and More Slow," in *New Dimensions 1: Fourteen Original Science Fiction Stories*, ed. Robert Silverberg (New York: Avon, 1973), 99–133, at 128.
11. James Lovelock, *Gaia: A New Look at Life on Earth* (Oxford: Oxford University Press, 1995), 10. Quoted in Thomas Pringle, "The Ecosystem Is an Apparatus: From Machinic Ecology to the Politics of Resilience," in *Machine*, ed. Thomas Pringle, Gertrud Koch, and Bernard Stiegler (Minneapolis: University of Minnesota Press and Menson Press, 2019), 49–103, at 77.

12. LeGuin, "Vaster than Empires and More Slow," 118.
13. Grass is considered in relation to the transformation of the prairies of the American Midwest into a place of industrial exhaustion; see Hannah Holleman, *Dust Bowls of Empire: Imperialism, Environmental Politics, and the Injustice of "Green" Capitalism* (New Haven, CT: Yale University Press, 2018); in relation to the European colonial project, see Richard Drayton, *Nature's Government: Science, Imperial Britain, and the "Improvement" of the World* (New Haven, CT: Yale University Press, 2000); and in relation to the longer cultural history of the metaphor of the green mantle, see Veronica della Dora. *The Mantle of the Earth: Genealogies of a Geographical Metaphor* (Chicago: University of Chicago Press, 2020).
14. See Chunglin Kwa, "Modeling the Grasslands," *Historical Studies in the Physical and Biological Sciences* 24, no. 1 (1993): 125–155. See also Tega Brain, "The Environment Is Not a System," *APRJA* 7, no. 1 (2018), https://researchvalues2018.wordpress.com/2017/12/20/tega-brain-the-environment-is-not-a-system/.
15. Pringle, "Ecosystem Is an Apparatus," 63.
16. Eugene P. Odum, "Energy Flow in Ecosystems: A Historical Review," *American Zoologist* 8, no. 1 (1968): 11–18. See Mark Glen Madison, "'Potatoes Made of Oil': Eugene and Howard Odum and the Origins and Limits of American Agroecology," *Environment and History* 3, no. 2 (1997): 209–238.
17. Also Lovelock's theorization of Gaia grew out of this period of mapping how surface levels interact with atmospheric chemistry in complex "informational self-regulation." See Pringle, "Ecosystem Is an Apparatus," 76. See also Giulia Rispoli, "Planetary Environing: The Biosphere and the Earth System," in *Environing Media*, ed. Adam Wickberg and Johan Gärdebo (London: Routledge, 2022), 54–74.
18. See Peder Anker, "The Ecological Colonization of Space," *Environmental History* 10, no. 2 (2005): 239–268. In this context it is good to note the science fiction novel by James Lovelock and Michael Allaby, *The Greening of Mars* (New York: Warner, 1985), and the work by Lynn Margulis on the (impossible) terraforming of planets such as Mars, for example in Lynn Margulis and Oona West, "Gaia and the Colonization of Mars," *GSA Today* 3, no. 11 (1993): 277–280, at 291.
19. Isaac Asimov, "Misbegotten Missionary (Green Patches)," *Galaxy Magazine*, November 1950, 34–47.
20. This was not a new concept introduced by Clements. As was fully acknowledged by the botanist in the historical sections in his books, the notion of plant formation had a well-known trajectory before his work. For more details on the history of this idea, see Frederic E. Clements, *Research Methods in Ecology* (Lincoln, NE: University Publishing Company, 1905), 2–4, and Tobey, *Saving the Prairies*, 76–109.
21. Clements, *Research Methods in Ecology*, 199.
22. Clements, *Research Methods in Ecology*, 199.

23. Boris Shoshitaishvili, "Is Our Planet Doubly Alive? Gaia, Globalization, and the Anthropocene's Planetary Superorganisms," *Anthropocene Review* 10, no. 2 (2022): 434–454.

24. Frederic E. Clements, *Plant Succession: An Analysis of the Development of Vegetation* (Washington, DC: Carnegie Institution of Washington, 1916), 3.

25. William Morton Wheeler, "The Ant Colony as an Organism," *Journal of Morphology* 22, no. 2 (2005): 307–325.

26. Jussi Parikka, *Insect Media: An Archaeology of Animals and Technology* (Minneapolis: University of Minnesota Press, 2010), 51–55. Kirchoff, "Myth of Frederic Clements's Mutualistic Organicism," 24.

27. Jan C. Smuts, *Holism and Evolution*, 2nd ed. (London: Macmillan, 1927), https://archive.org/stream/holismandevolutio32439mbp/holismandevolutio32439mbp_djvu.txt.

28. "This refusal is however far from meaning that I do not realise that various 'biomes,' the whole webs of life adjusted to particular complexes of environmental factors, are real 'wholes,' often highly integrated wholes, which are the living nuclei of systems in the sense of the physicist. Only I do not think they are properly described as 'organisms' (except in the 'organicist' sense). I prefer to regard them, together with the whole of the effective physical factors involved, simply as 'systems'"; Tansley, "Use and Abuse of Vegetational Concepts and Terms," 297.

29. On Tansley's critical stance to holism, as well as the context of racialized notions of "imperial ecology," see Peder Anker, *Imperial Ecology: Environmental Order in the British Empire, 1895–1945* (Cambridge, MA: Harvard University Press, 2001), 118–156.

30. Kirchoff, "Myth of Frederic Clements's Mutualistic Organicism," 8.

31. Frederic E. Clements and Victor E. Shelford, *Bio-Ecology* (New York: Wiley, 1939), 238–239; quoted in Kirchoff "Myth of Frederic Clements's Mutualistic Organicism," 9.

32. Clements and Shelford, *Bio-Ecology*, 271; quoted in Kirchoff, "Myth of Frederic Clements's Mutualistic Organicism," 9.

33. Kirchoff, "Myth of Frederic Clements's Mutualistic Organicism," 13–14.

34. Clements, *Research Methods in Ecology*, 270.

35. Clements, *Research Methods in Ecology*, 271.

36. Mary Louise Pratt, *Imperial Eyes: Travel Writing and Transculturation* (London: Routledge, 2008), 169–194.

37. Marie-Noëlle Bourguet, "Landscape with Numbers: Natural History, Travel and Instruments in the Late Eighteenth and Early Nineteenth Centuries," in *Instruments, Travel and Science. Itineraries of Precision from the Seventeenth to the Twentieth Century*, ed. Marie-Noëlle Bourguet, Christian Licoppe, and H. Otto Sibum (London: Routledge, 2002), 96–125, at 106.

38. Tobey, *Saving the Prairies*, 50.

39. Richard Hölzl, "Scientific Forestry in the Eighteenth and Nineteenth Centuries," *Science as Culture* 19, no. 4, (2010): 431–460.
40. Tobey, *Saving the Prairies*, 52.
41. Tobey, *Saving the Prairies*, 70.
42. Tobey, *Saving the Prairies*, 65.
43. Geoffrey Batchen, "Electricity Made Visible," in *New Media, Old Media: A History and Theory Reader*, ed. Wendy Hui Kyong Chun and Thomas Keenan (New York: Routledge, 2002), 27–44, at 28.
44. Batchen, "Electricity Made Visible," 28.
45. Batchen, "Electricity Made Visible," 29.
46. Batchen, "Electricity Made Visible," 29–30.
47. Batchen, "Electricity Made Visible," 30.
48. William Henry Fox Talbot, *The Pencil of Nature* (London: Longman, Brown, Green and Longmans, 1844; repr. 2010), 33.
49. Gilbert Simondon, *Individuation in Light of Notions of Form and Information*, vol. 1, trans. Taylor Adkins (Minneapolis: University of Minnesota Press, 2020), 246.
50. Oliver W. Holmes Sr., "The Stereoscope and the Stereograph," in *Classic Essays on Photography*, ed. Alan Trachtenberg (New Haven, CT: Leete's Island, 1980), 72–82, at 77, 78; quoted by John Durham Peters in "Space, Time, and Communication Theory," *Canadian Journal of Communication* 28, no. 4 (2003): 397–411, at 403.
51. On early photographic images and information abundance, referring to an "overcrowding of detail," see Mary Ann Doane, "Temporality, Storage, Legibility. Freud, Marey, and the Cinema," *Critical Inquiry* 22, no. 2 (1996): 313–343, at 328.
52. Clements, *Research Methods in Ecology*, 271.
53. Lorraine Daston and Peter Galison, *Objectivity* (New York: Zone, 2007).
54. Siegfried Zielinski, *Deep Time of the Media: Toward an Archaeology of Hearing and Seeing by Technical Means*, trans. Gloria Custance (Cambridge, MA: MIT Press, 2006), 215. Anecdotally, this use of statistics also features in W. E. B. Dubois's work; see Maria Farland, "W. E. B. DuBois, Anthropometric Science, and the Limits of Racial Uplift," *American Quarterly* 58, no. 4 (2006): 1017–1045.
55. On the role of statistics at the University of Nebraska, see Tobey, *Saving the Prairies*, 53–57; on their understanding of Darwin's work, see 69.
56. On the importance of counting populations in relation to the development of statistics, see Ian Hacking, *The Taming of Chance* (Cambridge: Cambridge University Press, 1990).
57. Tobey, *Saving the Prairies*, 74.
58. Tobey, *Saving the Prairies*, 55.
59. Kohler, *Landscapes and Labscapes*, 102.

60. Experimental practices were also tested, making the field into a lab: so-called denuded quadrats were "cleared of all vegetation by herbicides or excavation" so as to allow observation of "which plants invade or resettle vacant ground"; Kohler, *Landscapes and Labscapes*, 104. As Kohler notes, this was also an early form of modeling how environments might react after "fire, disease, or human interventions" (105).

61. Clements, *Research Methods in Ecology*, 188

62. Clements, *Research Methods in Ecology*, 188–196.

63. Clements, *Research Methods in Ecology*, 191. On the broader link between addressability, logistical media, and computation, see Ranjodh Singh Dhaliwal, "On Addressability, or What Even Is Computation?," *Critical Inquiry* 49, no. 1 (2022): 1–27.

64. Cindy Lin, "How Forests Became Data: The Remaking of Ground Truth in Indonesia," in *The Nature of Data: Infrastructures, Environments, Politics*, ed. Jenny Goldstein and Eric Host (Lincoln: University of Nebraska Press, 2022), 285–302.

65. Quoted in Shannon Mattern, "Tree Thinking," *Places*, September 2021, https://placesjournal.org/article/tree-thinking/.

66. Jennifer Gabrys, "Smart Forests and Data Practices: From the Internet of Trees to Planetary Governance," *Big Data & Society* 7, no. 1 (2020): 4. Mattern, "Tree Thinking."

67. See Birgit Schneider and Lynda Walsh, "The Politics of Zoom: Problems with Downscaling Climate Visualizations," *Geo: Geography and Environment* 6, no. 1 (2019): 1–11.

68. On these data ensembles in relation to Google Street View, see Ingrid Hoelzl and Rémi Marie, "Google Street View: Navigating the Operative Image," *Visual Studies* 29, no. 3 (2014): 261–271.

69. Steven Brumby, "Teaching the Cloud to See the Earth," TEDx Talks, October 15, 2015, YouTube video, 9:06, https://www.youtube.com/watch?v=VLze6W7SvyY.

70. Della Dora, *Mantle of the Earth*.

71. Current techniques devised to build forest and plantation inventories include automated tree crown detection, delineation, classification, and counting. See, for instance, Aishwarya Chandrasekaran, Guofan Shao, Songlin Fei, Zachary Miller, and Joseph Hupy, "Automated Inventory of Broadleaf Tree Plantations with UAS Imagery," *Remote Sensing* 14, no. 8 (2022): 1931.

72. Dhaliwal, "On Addressability, or What Even Is Computation."

73. May, "Logic of the Managerial Surface," 117.

74. Clements, *Research Methods in Ecology*, 274.

75. Tansley, "Use and Abuse of Vegetational Concepts and Terms."

76. Clements, *Research Methods*, 270.

77. Frederic E. Clements, *Plant Indicators: The Relation of Plant Communities to Process and Practice* (Washington, DC: Carnegie Institution of Washington, 1920), 3.

78. Clements, *Plant Indicators*, 3.

79. Nathaniel B. Ward, *On the Growth of Plants in Closely Glazed Cases*, 2nd ed. (London: John van Voorst, 1852), 14.

80. On a-signifying semiotics and signaletic matter, see Gary Genosko, *Félix Guattari: A Critical Introduction* (London: Pluto, 2009).

81. Kohler, *Landscapes and Labscapes*, 122. "In a general way, plants serve as measures of conditions wherever they grow. This is the principle that underlies the indicator value of native plants. It operates with equal force in the case of cultivated plants and crops of all sorts, though with these it must be recognized that the habitat has been artificially modified to some degree. Whenever a species is planted under new conditions of soil or climate, it serves as a kind of practical phytometer by comparison with its usual growth." Frederic E. Clements and Glenn W. Goldsmith, *The Phytometer Method in Ecology: The Plant and Community as Instruments* (Washington, DC: Carnegie Institution of Washington, 1924), 5.

82. Clements and Goldsmith, *Phytometer Method in Ecology*, 120; italics by the authors.

83. Gernot Böhme, "The Physiognomy of a Landscape" trans. Axel Häusler, in *Thermodynamic Interactions: Architectural Exploration into Material, Physiological and Territorial Atmospheres*, ed. Javier Garcia-German (New York: Actar, 2017), 219–234.

84. Orit Halpern, *Beautiful Data: A History of Vision and Reason since 1945* (Durham, NC: Duke University Press, 2014), 21.

85. Tansley distinguishes autogenic succession of plant formations from allogenic succession "in which the changes are brought about by external factors." Tansley, "Use and Abuse of Vegetational Concepts and Terms," 287.

86. Chris Salter, *Sensing Machines: How Sensors Shape Our Everyday Life* (Cambridge, MA: MIT Press, 2022), 15–40.

87. To refer to ecological, environmental surfaces as affordances that put into circulation all sorts of images, data, and actions is one form of continuing the idea of active surfaces. This aligns with J. J. Gibson's ecological notion of affordances—that material environments afford different capacities of perception, movement, and action—but it also brings into play the point that environmental surfaces are actively involved in their own interpretation, comparison, numbering, counting, and even imaging in ways that expand the repertoire of cultural techniques to ecological aspects of technical systems. Continuing our previous chapter's theme of preparing ground surfaces for machine vision, we can see a similar theme about environment surfaces as dynamic in this fundamental sense of affordance. See James J. Gibson, *The Ecological Approach to Visual Perception* (Hillsdale, NJ: Lawrence Erlbaum, 1986). On Gibson and affordance in media theory, see Parikka, *Insect Media*, passim.

88. Pietrusko, "Ground Cover," 16.

89. Pietrusko, "Ground Cover," 16.

90. Pietrusko, "Ground Cover," 16.
91. Pietrusko, "Ground Cover," 14
92. Pietrusko, "Ground Cover," 18.
93. Pietrusko, "Ground Cover," 16.
94. Pietrusko, "Ground Cover," 17.
95. Cf. Cubitt on Ansel Adams's zoning in Seán Cubitt, *The Practice of Light: A Genealogy of Visual Technologies from Prints to Pixels* (Cambridge, MA: MIT Press, 2014), 91–95.
96. Jacob Darwin Hamblin, *Arming Mother Nature: The Birth of Catastrophic Environmentalism* (Oxford: Oxford University Press, 2013); David Zierler, *The Invention of Ecocide: Agent Orange, Vietnam, and the Scientists Who Changed the Way We Think about the Environment* (Athens: University of Georgia Press, 2011).
97. Kristen Jordan, "Advanced Plant Technologies," *DARPA*, accessed August 8, 2023, https://www.darpa.mil/program/advanced-plant-technologies.
98. Clements, quoted in Kohler, *Labscape and Landscape*, 122.
99. Jordan, "Advanced Plant Technologies."
100. Blake Bextine, the DARPA Program Manager for APT, quoted in "Nature's Silent Sentinels Could Help Detect Security Threats," *DARPA*, November 17, 2017, https://www.darpa.mil/news-events/2017-11-17.
101. Kohler, *Labscape and Landscape*, 118–122.
102. Jennifer Gabrys, *Program Earth: Environmental Sensing Technology and the Making of a Computational Planet* (Minneapolis: University of Minnesota Press, 2016), 15.
103. Gabrys, *Program Earth*, 43.

CHAPTER 7

1. Another realm of light where images are literally diving deeper are oceans and other hydric formations. Algae and other marine vegetation and phenomena such as uncontrolled blooms, species invasions of different kinds and the resulting eutrophication of waters parallel the spread of monitoring infrastructure such as specific satellites and machine learning computation. On this topic see Melody Jue, "'Pixels May Lose Kelp Canopy': The Photomosaic as Epistemic Figure for the Satellite Mapping and Modeling of Seaweeds," *Media+Environment* 3, no. 2 (2021), https://mediaenviron.org/article/21261-pixels-may-lose-kelp-canopy-the-photomosaic-as-epistemic-figure-for-the-satellite-mapping-and-modeling-of-seaweeds. We also want to point to Alejandro Limpo's ongoing PhD project (University of Southampton) on visual culture and cultural techniques of oceans and thank him for his research assistance in locating relevant sources.
2. See Robert P. Harrison, *Forests: The Shadow of Civilization* (Chicago: University of Chicago Press, 1993), 197–243.

3. Jennifer Gabrys, "Smart Forests and Data Practices: From the Internet of Trees to Planetary Governance," *Big Data & Society* 7, no. 1 (2020); Rosetta S. Elkin, *Plant Life: The Entangled Politics of Afforestation* (Minneapolis: University of Minnesota Press, 2022). See also Shannon Mattern, "Tree-Thinking," *Places*, September 2021, https://placesjournal.org/article/tree-thinking/.
4. For a wide account of cases of "environcides," including receding forests all around the world ranging from the sixteenth to the early twentieth centuries, see Emmanuel Kreike, *Scorched Earth: Environmental Warfare as a Crime against Humanity and Nature* (Princeton, NJ: Princeton University Press, 2021). On the political and aesthetical dimension of the forest, away from the usual romanticization of its shadowlands, see Matthew Fuller and Olga Goriunova, *Bleak Joys: Aesthetics of Ecology and Impossibility* (Minneapolis: University of Minnesota Press, 2019), 121–153.
5. Gabrys. "Smart Forests and Data Practices."
6. The limiting zones between ecosystems of different kinds display characteristics similar to the transitions of plant formations that we discuss in chapter 6. They are thus fertile regions in relation to the data-gathering processes they give rise to. On the cultures of data in between the Amazon rainforest and the savanna, for example, see Bruno Latour, "Circulating Reference: Sampling the Soil in the Amazon Forest," in *Pandora's Hope: Essays on the Reality of Science Studies*, ed. Bruno Latour (Cambridge, MA: Harvard University Press, 1999), 24–79.
7. The green cascade in the original phrase refers to vegetal matter, not light: "V. Panfilov says that tropical forests resemble a gigantic green cascade frozen in its downfall." The quote is found in Andrei V. Lapo, *Traces of Bygone Biospheres* (Moscow: MIR, 1982), 134.
8. Ute Holl, *Cinema, Trance and Cybernetics* (Amsterdam: Amsterdam University Press, 2017), 19.
9. Fuller and Goriunova, *Bleak Joys*, 123.
10. For an analysis of the capacity of immersive environments—such as Anouk de Clercq's LiDAR imaging of urbanscapes as similar to forest point clouds—to create effective simulations of altered states of consciousness, centered on the use of darkness, see Martine Beugnet and Lily Hibberd, "Cinematic Darkness: Dreaming across Film and Immersive Digital Media," *AN-ICON. Studies in Environmental Images* 1, no. 1 (2022): 129–152.
11. Giuliana Bruno, *Atmospheres of Projection: Environmentality in Art and Screen Media* (Chicago: University of Chicago Press, 2022), 38.
12. Giuliana Bruno, *Surface: Matters of Aesthetics, Materiality, and Media* (Chicago: University of Chicago Press, 2014), 67.
13. Bruno, *Surface*. The quotation is from Georges Didi-Huberman, "The Fable of the Place," in *James Turrell: The Other Horizon*, ed. Peter Noever (Vienna: MAK and Hatje Cantz, 2001), 45–56.
14. Bruno, *Surface*, 67.

15. Bruno, *Surface*, 67.

16. Eduardo Kohn, *How Forests Think: Toward an Anthropology beyond the Human* (Berkeley: University of California Press, 2013), 78, 9.

17. Marcel Minnaert, *The Nature of Light and Colour in the Open Air* (Mineola, NY: Dover, 1954), 1.

18. Mabel L. Todd, *Total Eclipses of the Sun* (Boston, MA: Roberts Brothers, 1894), 20. For a natural (media) history of light from technical apparatuses to such natural cinema, see Amédée Guillemin, *Le Monde physique*, vol. 2, *La lumière* (Paris: Hachette, 1882).

19. Matthew Fuller and Eyal Weizman, *Investigative Aesthetics: Conflicts and Commons in the Politics of Truth* (London: Verso, 2021), 50.

20. Fuller and Weizman, *Investigative Aesthetics*, 50.

21. Margaret C. Anderson and Edward E. Miller, "Forest Cover as a Solar Camera: Penumbral Effects in Plant Canopies," *Journal of Applied Ecology* 11, no. 2 (1974): 691–697.

22. See Nancy Anderson and Michael R. Dietrich, eds., *The Educated Eye: Visual Culture and Pedagogy in the Life Sciences* (Hanover, NH: Dartmouth College Press, 2012).

23. John A. Endler, "The Color of Light in Forests and Its Implications," *Ecological Monographs* 63, no. 1 (1993): 1–27, at 1.

24. Margaret C. Anderson, "Studies of the Woodland Light Climate: I. The Photographic Computation of Light Conditions," *Journal of Ecology* 52, no. 1 (1964): 27–41.

25. Eva Horn, "Air as Medium," *Grey Room*, December 1, 2018, 6–25, at 10–13, 12.

26. Nancy L. Peluso and Peter Vandergeest, "Political Ecologies of War and Forests: Counterinsurgencies and the Making of National Natures," *Annals of the Association of American Geographers* 101, no. 3 (2011): 587–608.

27. For a classic reference on state-driven scientific forestry, see James C. Scott, *Seeing Like a State: How Certain Schemes to Improve the Human Condition Have Failed* (New Haven, CT: Yale University Press, 1999), 11–22. For a more extended approach, see also Richard Hölzl, "Historicizing Sustainability: German Scientific Forestry in the Eighteenth and Nineteenth Centuries," *Science as Culture* 19, no. 4 (2010): 431–460.

28. Anderson, "Studies of the Woodland Light Climate," 27. Anderson emphasizes the importance that Wiesner gave to absolute values of Lichtgenuss when comparing different sites: "Wiesner was clear that comparative measurements of relative 'Lichtgenuss' (instantaneous diffuse site factor) of plant communities from different climates and latitudes could not be used as a basis for comparison of light conditions. Comparison could only be made between absolute, not relative measurements, since the light conditions in the open vary widely with climate and latitude. This point seems subsequently to have been generally overlooked" (39).

29. A summary of this experimentation through different photographic techniques can be followed in G. C. Evans and D. E. Coombe, "Hemisperical and Woodland Canopy Photography and the Light Climate," *Journal of Ecology* 47, no. 1 (1959): 103–113, at 103–104; and in Anderson, "Studies of the Woodland Light Climate," 31–32.

30. For a wider discussion on the relation between addressability and computation, see Ranjodh Singh Dhaliwal. "On Addressability, or What Even Is Computation?," *Critical Inquiry* 49, no. 1 (2022): 1–27. On grids and the "subordination of physical space to computational control," see Jeffrey Moro, "Grid Techniques for a Planet in Crisis: The Infrastructures of Weather Prediction," *Amodern* 9 (April 2020), https://amodern.net/article/grid-techniques/. See also Jussi Parikka, *Operational Images: From the Visual to the Invisual* (Minneapolis: University of Minnesota Press, 2023), 51–56.

31. Anderson, "Studies of the Woodland Light Climate," 33. The method consisted of counting the cells in the grid after classifying them into five classes, depending of leaves-to-sky ratio: "The second complication is that the scale of details of canopy structure is much finer than the scale of the grid. It is necessary to classify the segments of the grid into five classes: entirely unobstructed, less than 33% obstructed, 33–66% obstructed, 66–90% obstructed, and more than 90% obstructed. These classes are then taken as contributing 100%, 75%, 50%, 25% and 0% of the possible illuminance to the total" (34). The similarities of this method with the ones on plant geography discussed in the last chapter are evident.

32. Anderson, "Studies of the Woodland Light Climate," 39.

33. The HemiView system is supplied by Delta-T. On smartphone hemispherical photography, see Simone Bianchi, Christine Cahalan, Sophie Hale, and James Michael Gibbons, "Rapid Assessment of Forest Canopy and Light Regime Using Smartphone Hemispherical Photography," *Ecology and Evolution* 7, no. 24 (2017): 10556–10566.

34. The use of lidar features in many forms of contemporary art practice. Works would include Anouk De Clearcq's *Thing*, and ScanLAB's *Equirectangular Landscapes 05: Nevada Falls (After Muybridge)*. See, respectively, Beugnet and Hibbert, "Cinematic Darkness," and Jussi Parikka, "On Seeing Where There's Nothing to See: Practices of Light beyond Photography," in *Photography Off the Scale: Technologies and Theories of the Mass Image*, ed. Tomáš Dvořák and Jussi Parikka (Edinburgh: Edinburgh University Press, 2021), 185–210.

35. On operational images and "operations other than war" as they feature in different epistemic and operational imaging practices, see Parikka, *Operational Images*.

36. On the details of his invention, see Robin Hill, "A Lens for Whole Sky Photographs," *Quarterly Journal of the Royal Meteorological Society* 50, no. 211 (1924): 227–235.

37. See Goddard's account of this research in George W. Goddard, "I Looked Back, and He Was Smiling," *University of Rochester Library Bulletin* 26, no. 3 (1971), https://rbscp.lib.rochester.edu/3595.

38. This was the average intensity of such photoflash bombs once their production had been industrialized. The Rochester cartridge contained 90 pounds of flashlight power, approximately the same amount as the later ones. For more detailed information on these bombs of lights, see *Department of the Army Technical Manual TM 9-1385-51: Identification of Ammunition (Conventional) for Explosive Ordnance Disposal* (Washington, DC: Headquarters, Department of the US Army: 1967), 72.11–72.16.
39. Goddard, "I Looked Back."
40. On lidar as a version of flash and active sensing, see Parikka, "On Seeing Where There's Nothing to See."
41. Peter Sloterdijk, *Terror from the Air*, trans. Amy Patton and Steve Corcoran (Los Angeles, CA: Semiotext(e), 2009), 9–46.
42. Richard A. Ruth. "The Secret of Seeing Charlie in the Dark: The Starlight Scope, Techno-Anxiety, and the Spectral Mediation of the Enemy in the Vietnam War," *Vulcan* 5, no. 1 (2017): 64–88, at 70.
43. Ruth. "Secret of Seeing Charlie in the Dark," 68.
44. Ian G. R. Shaw. "Scorched Atmospheres: The Violent Geographies of the Vietnam War and the Rise of Drone Warfare," *Annals of the American Association of Geographers* 106, no. 3 (2016): 688–704, at 689, 694.
45. Yuriko Furuhata, *Climatic Media: Transpacific Experiments in Atmospheric Control* (Durham, NC: Duke University Press, 2022), 35.
46. Deborah Shapley, "Weather Warfare: Pentagon Concedes 7-Year Vietnam Effort," *Science* 184, no. 4141 (1974): 1059–1061, at 1060. See also Ryan Bishop, "Military Meteorology," in *Words of Weather*, ed. Daphne Dragona and Jussi Parikka (Athens: Onassis Stegi), 129–134.
47. Shapley, "Weather Warfare," 1061.
48. Shaw, "Scorched Atmospheres," 694.
49. Shaw, "Scorched Atmospheres," 694.
50. Shaw, "Scorched Atmospheres," 693.
51. Furuhata, *Climatic Media*, 2; see also 31–38.
52. Charles Vernon Boys, *Soap-Bubbles: Their Colours and the Forces which Mould Them* (New York: E. S. Gorham, 1912), 128. On soap bubbles and their use as models for different scientific phenomena from the late decades of the nineteenth to the early twentieth century, see Simon Schaffer, "A Science Whose Business Is Bursting: Soap Bubbles as Commodities in Classical Physics," in *Things that Talk: Object Lessons from Art and Science*, ed. Lorraine Daston (New York: Zone, 2004), 147–194.
53. Paul N. Edwards, *A Vast Machine: Computer Models, Climate Data, and the Politics of Global Warming*, ed. Geoffrey C. Bowker (Cambridge, MA: MIT Press, 2013), 418–419.
54. Edwards, *Vast Machine*.

55. Jane Hutton, *Reciprocal Landscapes: Stories of Material Movements* (Abingdon: Routledge, 2020), 11.

56. See Kreike, *Scorched Earth*.

57. On the discussions inside the US Army about the counterinsurgency actions to deal with this undetectable enemy, see David Zierler, *The Invention of Ecocide: Agent Orange, Vietnam, and the Scientists Who Changed the Way We Think about the Environment* (Athens: University of Georgia Press, 2011), 56–58.

58. Hannah Meszaros Martin. "'Defoliating the World': Ecocide, Visual Evidence and 'Earthly Memory,'" *Third Text* 32, no. 2–3 (2018): 230–253. On ecocide and forensic methods, see also Nabil Ahmed, "Proof of Ecocide: Towards a Forensic Practice for the Proposed International Crime against the Environment," *Forensic Archaeology, Anthropology and Ecology* 1, no. 2, (2019): 139–147.

59. The image complex, or the architectural image complex, is a term borrowed from forensic architecture. "What we refer to as the architectural image complex is a method of assembling image evidence in a spatial environment. The architectural image complex can function as an optical device that allows the viewer to see the scene of the crime as a set of relations between images in time and space. It can also be used as a navigational device to help move between images, exploring a space that is at once virtual and photographic. Essentially, it makes manifest the necessity for composing evidence that is simultaneously material, media-based, and testimonial. The architectural image complex thus replaces both the thematic classification system of archives and the linear transition between images in before-and after montages." Eyal Weizman, *Forensic Architecture: Violence at the Threshold of Detectability* (New York: Zone, 2017), 100.

60. Pujita Guha, "Seeding the Forest," *Cultural Politics* 19, no. 2 (2023). On herbicidal warfare in Vietnam, see Zierler, *Invention of Ecocide*. On a critical approach to the US Air Force's "electronic barrier," focusing on its reliance on the persistence of manual operation work, see David Young, "Sensors, Interpreters, Analysts: Operating the 'Electronic Barrier' during the Vietnam War," *Digital War* 2 (2021): 51–63.

61. Ryan Bishop, "Smart Dust and Remote Sensing: The Political Subject in Autonomous Systems," *Cultural Politics* 11, no. 1 (2015): 100–109.

62. Gabrys. "Smart Forests and Data Practices."

63. Gabrys, "Smart Forests and Data Practices," 7.

64. See Jennifer Gabrys, "Becoming Planetary," *e-flux*, October 2018, https://www.e-flux.com/architecture/accumulation/217051/becoming-planetary/. On the concept of planetarity, see Gayatri Chakravorty Spivak, *Death of a Discipline* (New York: Columbia University Press, 2003), 71–102. On its relation with the Cold War's ecologies of remote sensing, see Elizabeth DeLoughrey, "Satellite Planetarity and the Ends of the Earth," *Public Culture* 26, no. 2 (2014): 257–280. See also John Beck and Ryan Bishop, eds., *Cold War Legacies: Systems, Theory, Aesthetics* (Edinburgh: Edinburgh University Press, 2016).

65. Fuller and Goriunova, *Bleak Joys*, 152. See also Mattern, "Tree-Thinking."

66. See Spivak, *Death of a Discipline*, 71–102.

67. Spivak, *Death of a Discipline*, 151.

68. Paulo Tavares, "The Geological Imperative: On the Political Ecology of Amazonia's Deep History," in *Architecture in the Anthropocene: Encounters Among Design, Deep Time, Science and Philosophy*, ed. Etienne Turpin (London: Open Humanities Press, 2013), 209–240. See also Ryan Bishop, "Project 'Transparent Earth' and the Autoscopy of Aerial Targeting: The Visual Geopolitics of the Underground," *Theory, Culture & Society* 28, no. 7–8 (2011): 270–286.

69. Geoff Manaugh, "Geomedia, or What Lies Below," *BLDGBLOG* (blog), December 31, 2020, https://bldgblog.com/2020/12/geomedia-or-what-lies-below/.

70. Paulo Tavares, "In the Forest Ruins," *e-flux*, December 2016, http://www.e-flux.com/architecture/superhumanity/68688/in-the-forest-ruins/.

71. Tavares, "In the Forest Ruins."

72. On other indigenous practices of mapping in relation to current design and technological culture, see Shannon Mattern, "Mapping's Intelligent Agents," *Places Journal*, September 2017, https://placesjournal.org/article/mappings-intelligent-agents/.

73. Keller Easterling, *Medium Design* (London: Verso, 2021), viii–ix. There is also an interesting link between such recent notions as environmental media and medium design with Serres's natural contract. See Michel Serres, *The Natural Contract*, trans. Elizabeth MacArthur and William Paulson (Ann Arbor: University of Michigan Press, 1995).

74. Here we are paraphrasing Lapo, *Traces of Bygone Biospheres*, a similar recursive and archaeological take on the scale of the planet, which is discussed also in chapter 3.

75. Rob Nixon, "The Less Selfish Gene," *Environmental Humanities* 13, no. 2 (2021): 348–371, at 353.

76. For an argument about forests as accomplishers of the industrial fantasy of locally produced biotech products and services, see, for example, Elliot Hershberg, "Atoms Are Local," *Century of Biology* (blog), November 7, 2022, https://centuryofbio.substack.com/p/atoms-are-local.

77. See also Lukáš Likavčan, *Introduction to Comparative Planetology* (Moscow: Strelka, 2019).

78. Gabrielle Hecht, "Interscalar Vehicles for an African Anthropocene: On Waste, Temporality, and Violence," *Cultural Anthropology* 33, no. 1 (2018): 109–141.

79. Eliza Rose, "Cold-War Cabin Ecologies: Soviet-American Biospheric Thinking," *Science Fiction Studies* 49, no. 2 (2022), 267–287; Peder Anker, *From Bauhaus to Ecohouse: A History of Ecological Design* (Baton Rouge: Louisiana State University Press, 2010).

80. This was the case also when we discussed how the scale of empires at the end of the nineteenth century brought in models of planetary phenomena that reshaped the planet itself.

81. Roger Stahl, "Becoming Bombs: 3D Animated Satellite Imagery and the Weaponization of the Civic Eye," *Mediatropes* 2, no. 2 (2010): 65–93, at 86.

82. For a different perspective of this metaphor in relation to the infrastructure of fashion, following the work of the studio Unknown Fields Division, see Jussi Parikka, "Folds of Fashion: *Unravelled* and the Planetary Surface," in *Surface and Apparition: The Immateriality of Modern Surface*, ed. Yeseung Lee (London: Bloomsbury, 2020), 19–36. See also Veronica della Dora, *The Mantle of the Earth: Genealogies of a Geographical Metaphor* (Chicago: University of Chicago Press, 2020).

83. Simryn Gill and Michael Taussig, *Becoming Palm*, ed. Ute Meta Bauer (Amsterdam: Sternberg, 2017), 11.

84. Jussi Parikka and Abelardo Gil-Fournier, "An Ecoaesthetic of Vegetal Surfaces: On Seed, Image, Ground as Soft Montage," *Journal of Visual Art Practice* 20, no. 1–2 (2021): 16–30.

85. The work of forensic anthropologist Margaret Cox on living surfaces, tracing mass graves, and its elaboration by artist Jananne al-Ani are examples of this. See Caren Kaplan, *Aerial Aftermaths: Wartime from Above* (Durham, NC: Duke University Press, 2018), 180–206; Margaret Cox, Ambika Flavel, Ian Hanson, Joanna Laver, and Roland Wessling, eds., *The Scientific Investigation of Mass Graves: Towards Protocols and Standard Operating Procedures* (Cambridge: Cambridge University Press, 2008).

86. Susan Schuppli, *Material Witness: Media, Forensics, Evidence* (Cambridge, MA: MIT Press, 2020), 257.

87. Wietske Maas, "The Corruption of the Eye: On Photogenesis and Self-Growing Images," *e-flux*, September 1, 2015, http://supercommunity.e-flux.com/texts/the-corruption-of-the-eye-on-photogenesis-and-self-growing-images/. Emanuele Coccia explores in depth such an approach in his essay on the "metaphysics of mixture": Emanuele Coccia, *The Life of Plants: A Metaphysics of Mixture*, trans. Dylan J. Montanari (Medford, MA: Polity, 2019).

88. In this, we are following the footsteps of the eco-aesthetic take by Seán Cubitt. See Cubitt, *Finite Media: Environmental Implications of Digital Technologies* (Durham, NC: Duke University Press, 2016).

89. Irmgard Emmelhainz, "Conditions of Visuality under the Anthropocene and Images of the Anthropocene to Come," *e-flux* 63 (March 2015), http://www.e-flux.com/journal/63/60882/conditions-of-visuality-under-the-anthropocene-and-images-of-the-anthropocene-to-come/. From the perspective of our book, the experimental film practices discussed in her essay—such as North American structural film in the 1960s and 1970s, dealing with forms of vision beyond a humanist-centered view—can be expanded to embrace larger circuits and scales of time and duration, which can even be taken as bioinfrastructure, even social, architectural, and media formations beyond solarity. From this expanded perspective, Emmelhainz's suggestion of reading anew Jean-Luc Goddard's prompt to understanding images as entities aiming not to show, know, or possess, but

as embodying a desire to see, emerges as an environmentally sound statement. On media formations beyond solarity, see Puig de la Bellacasa, *Matters of Care: Speculative Ethics in More than Human Worlds* (Minneapolis, MN: University of Minnesota Press, 2017); Nicole Starosielski, "Beyond the Sun: Embedded Solarities and Agricultural Practice," *South Atlantic Quarterly* 120, no. 1 (2021): 13–24. On plants as dead labor, see Nadia Bozak, *The Cinematic Footprint: Lights, Camera, Natural Resources* (New Brunswick, NJ: Rutgers University Press, 2012); Luce Lebart, "La photographie d'origine végétale," in *Puissance du végétal et cinéma animiste: La vitalité révélée par la technique*, ed. Teresa Castro, Perig Pitrou, and Marie Rebecchi (Paris: Les Presses du Réel, 2020), 119–127.

90. Walter Benjamin, *The Arcades Project*, trans. Howard Eiland and Kevin McLaughlin (Cambridge, MA: Belknap Press of Harvard University Press, 1996), 390 (Kla, 3).

91. The original quote is "the world and its potential for becoming are recursively remade in each meeting"; Karen Barad, *Meeting the Universe Halfway: Quantum Physics and the Entanglement of Matter and Meaning* (Durham, NC: Duke University Press, 2007), x.

92. Barad, *Meeting the Universe Halfway*, 153.

INDEX